THE VEGETATION OF EUROPE

This is a volume in the Arno Press collection

HISTORY OF ECOLOGY

Advisory Editor
Frank N. Egerton III

Editorial Board
John F. Lussenhop
Robert P. McIntosh

*See last pages of this volume for a
complete list of titles.*

THE

VEGETATION OF EUROPE

ITS

CONDITIONS AND CAUSES

ARTHUR HENFREY

ARNO PRESS

A New York Times Company

New York / 1977

Editorial Supervision: LUCILLE MAIORCA

———◆———

Reprint Edition 1977 by Arno Press Inc.

Reprinted from a copy in
  The University of Pennsylvania Library

HISTORY OF ECOLOGY
ISBN for complete set: 0-405-10369-7
See last pages of this volume for titles.

Manufactured in the United States of America

Publisher's Note: The maps have been reproduced
in black and white in this edition.

———◆———

Library of Congress Cataloging in Publication Data

Henfrey, Arthur, 1819-1859.
    The vegetation of Europe.

    (History of ecology)
    Reprint of the 1852 ed. published by J. Van Voorst,
London, which was issued as v.1 of Outlines of the
natural history of Europe.
    1. Botany--Europe--Ecology. I. Title. II. Series.
III. Series: Outlines of the natural history of Europe
; v.1.
QK281.H44 1977   581.5'094   77-74227
ISBN 0-405-10397-2

OUTLINES

OF

THE NATURAL HISTORY OF EUROPE.

THE

VEGETATION OF EUROPE,

ITS

CONDITIONS AND CAUSES.

BY

ARTHUR HENFREY, F.L.S.,

&c. &c.

LONDON:
JOHN VAN VOORST, PATERNOSTER ROW.
MDCCCLII.

CONTENTS.

*** The Explanation of the Map will be found at page 75.

THE VEGETATION OF EUROPE.

———◆———

CHAPTER I.

INTRODUCTORY.

THE pages of the book of Nature offer a vast variety of characters for our perusal, but of all we find there inscribed, none surpass in beauty of form or the interest of their revelations those presented by vegetable life. Mountain and valley, flood and lake, plain and undulating hill, may give the bolder features of a landscape, but dark and cheerless must the grandest combination of forms appear where the eye can find no green resting-place ; gloomy and repulsive the scene where no trace of vegetation, telling with its varying hue the tale of life and change, breaks the dull monotony of the stark masses of the earth's crust. It is difficult, indeed, to those who are without the actual experience, to picture in the mind those desert tracts which do actually exist upon the globe, where the burning sun sears out as it were the ordinary covering of the soil ; or those barren rocky shores, clothed

but by a few lichens, where the inhospitable climate
refuses a resting-place even to a blade of grass. We are
connected with the vegetable world by so many ties, of
pleasure, interest, and necessity, that we commonly regard
its existence as a matter of course, and seldom pause to
consider how and why it is, but merely direct our attention to those of its peculiarities which relate to its useful
qualities, or which lend to it the manifold charms which
delight our gaze in natural scenery. From the very
infancy of our race the influence of vegetation upon the
moral feelings has been recognized, poets have dwelt
upon it in all ages, and scarcely a striking form or commonly recurring kind of plant is without its real or
fanciful associations. Waving corn-fields! even the bare
mention of them seems to raise a vision of peace, plenty
and contentment : traversing the woodland path we cast
awhile the cares that press upon us in the busy haunts
of congregated man, and share the freedom and independence of the unrestrained life around ; or in the deep
and silent solitude of the black pine forest, we feel re-
vive within us that superstitious awe that gave birth to
the strange traditions of our northern ancestors. No
temperament at all awake to the influence of external
nature can escape the depressing influence of the low
swampy plain, where among plashing water-courses the
" cluster'd marish-mosses " creep, and amid the rustling
reeds

> " Willows whiten, aspens quiver,
> Little breezes dusk and shiver."

Few behold unmoved the brilliant tints of the spring
buds, fringing the spray of every re-awakening tree like
a joyous decoration celebrating the return of warmth and

active life, or the graduated hues of autumn's garb ;—they typify to us too plainly the changeful course of our own existence not to arouse something more than a passing feeling of wonder or admiration.

And descending to particulars, we recognize in common epithets, endless examples of the association of moral qualities with certain plants, as the 'sturdy oak,' the 'modest daisy,' the 'melancholy poplar,' the 'humble celandine,' and the like ; every page of descriptive poetry or imaginative prose, in our own and in all other literatures, will furnish instances.

In regard, again, to their mere physical beauty, the more conspicuous tribes of the vegetable world, the trees, have furnished the principal materials for the development of one great branch of Art. The infinite variety of forms, the delicacy and multiplicity of lines, surfaces and colours, the changing aspect in different seasons and under varying lights, have presented an inexhaustible field of study to the painter, and one which has attracted and rewarded the most distinguished masters.

But poets had sung, and painters had turned the powers of their art to the representation of the varied landscapes of different regions, commerce had heaped the products of every latitude in the market-places of civilized nations, and travellers had raised the wonder and curiosity alike of the learned and the ignorant with tales of the strange forms replacing our familiar plants in foreign climes ; yet it was not until about the beginning of the present century that science made any general or systematic inquiries in this direction, and Botanical Geography, as a special subject, owes its origin to the labours of living philosophers, Robert Brown, Alexander von Humboldt, Schouw and others, who during the last fifty

years have attempted to arrange the facts and develope
the laws of these phænomena.

Not that the facts lay out of common view, for they
are among the most striking and prominent in the whole
range of knowledge. The contrasts and diversities among
the characteristic vegetations of different lands force
themselves upon the most superficial observer. Since
the earliest period of which we have record, the peculiarity of certain plants to certain countries or regions
has attracted observation, and the narratives of the earlier
navigators of European nations are full of glowing pictures
of the treasures unfolded to them in the more favoured
climes to which they penetrated. The tropics were depicted as earthly paradises, in which

" Droops the heavy-blossom'd bower, hangs the heavy-fruited tree,
   Summer isles of Eden lying in dark-purple spheres of sea ; "

and in which man lies idly waiting while nature pours out
at his feet the rich harvest of luxurious fruits, unknown
in our temperate regions, where the ever-recurring check
periodically arrests the forces of vegetation, and the less
favouring climate compels him to the labours of the field,
yielding to his toil and unremitting care a limited and
frugal recompense. Travellers had told, too, that in the
far north even this partial bounty is denied ; that man is
cut off altogether from that vegetable food which is
lavished in profusion at his slightest demands beneath a
warmer sky, while his companion the reindeer scrapes a
scanty repast from beneath the snowy covering of the soil.

"The carpet of flowers and of verdure spread over
the naked crust of our planet is unequally woven ; it is
thicker where the sun rises high in the ever-cloudless
heavens, and thinner toward the poles, in the less happy

climes where returning frosts often destroy the opening buds of spring or the ripening fruits of autumn."

But even in a smaller compass, on a smaller field, striking differences occur, and facts familiar to every educated person mark the existence of some regulating influence even within the limits of the smallest of the continents of the world. We cultivate the grape in England, but it is only in favoured spots, and then even not with certainty, that it will ripen properly in the open air; yet but a little further south it is so much at home, that it yields one of the necessaries of life to the entire population. Oranges will ripen on the other side of the Alps, but not on this. For those cereal grains, those corn-plants furnishing the principal portion of the food of man, we find distinct lines of demarcation extending across Europe, beyond which, northward, each kind ceases to be capable of ripening its seed. Of trees we know that certain kinds will flourish and form fruits at points far north, where others are arrested by the cold; the firs, for instance, exclusively constitute the most northern woods of Scandinavia, while the dwarf palm, a representative of tropical climates, maintains its footing even so far into the temperate region as Italy and the southern confines of France.

Again, as indeed must be perceptible to every one who has visited mountainous countries, vegetation alters in its characters at different elevations, and it has been shown that these variations correspond to those which are observed on the level plains in proceeding from the south towards the north; the increased severity of the climate of the higher localities acting exactly in the same way as the colder climate of the regions lying further from the equator.

Such facts as these, obvious as they appear to be, remained unconnected and unaccounted for until recent times, or differences of heat and cold were supposed to be sufficient to explain them. But when a more searching inquiry arose, and when the vague ideas respecting the influence of heat came to be systematically investigated, it was found that there were other facts and that other causes were at work, the existence of which had not previously been suspected. In the first place it was seen, that mere degree of latitude will not indicate the temperature of a climate; that the temperature, the average heat and cold, do not alone constitute the climate properly so-called, but that humidity, exposure to prevailing winds and many other influences conjoin to produce the atmospheric conditions powerfully affecting vegetation. The chemical and physical conditions of the soil were found to require investigation, in order to the explanation of facts otherwise anomalous; and finally it has been discovered that the particular constitutions of the individual species of plants must be studied, if we would rightly understand the causes which give the peculiar characters to the vegetation of different lands.

And after all these points have been considered, there is still a residuum of phænomena which they totally fail to account for; we are in possession of another series of facts which require to be explained by a wholly different set of causes, as will appear from the following observations.

When we compare the floras, that is, the lists of native plants, of two countries closely alike in physical conditions, we generally find a difference resulting from the absence of certain kinds in one which exist in the other, and *vice versâ*; moreover, these may be kinds, which when introduced into the country where they were wanting will flourish there with a luxuriance equalling that in their native habitation. This shows it is not the physical or external conditions which have prevented their growth, and we therefore ask why were not they present at first in both? This leads to the inquiry into the details of the peculiarities of the vegetation of different countries, and we then become aware that there is some law presiding over the distribution of plants which causes the appearance of particular species arbitrarily, if we may so say it, in particular places; and following the clue which this affords, we arrive at the conclusion that countries have become populated with plants partly by the spreading of some special kinds from centres within those countries where they were originally exclusively created, and while these have spread outward into the neighbouring regions, colonists from like centres lying in the surrounding countries have invaded and become intermingled with the indigenous inhabitants. The modes in which these processes have gone on, the details of the migrations, and similar particulars, are matters of much debate and discussion, and require great care in their determination; but it is now generally admitted that such centres of creation do exist, and thus we have here, side by side with the climatic and other physical influences, a second and totally different set of conditions, which must be thoroughly investigated before we can clearly understand the manner in which the vegetable inhabitants of the world have acquired their present positions and relations toward each other. When we have to deal with a great extent of the earth's surface, the phænomena presented by this branch of the subject are very striking, sufficiently so to have enabled M. Schouw, one of the most distinguished geographical botanists, to lay down regions on the map of the world, in which particular forms are so predominant as to give a peculiar character to the vegetation, these peculiarities not pointing at a difference of climate, but often indicating rather a resemblance, where certain tribes of plants are *represented*, as it were, by other tribes quite distinct in structure, but agreeing very closely in the habits which place them under the influence of the external physical agencies. In a confined space, within the limits of a region so little varied as the continent of Europe, it is not easy to illustrate this part of the subject very clearly to those who are unacquainted with the more minute distinctions between different plants; but facts, and most marked ones, exist in abundance, and to the botanist, the results here are, if possible, even more convincing, because the region has been so much more thoroughly investigated. In Europe every country has its active naturalists; our own country, Germany, France and Scandinavia have zealous botanists in almost every county, province or department, examining with curious eye every tract within their reach, and eager to detect any new fact respecting the distribution of the plants of their vicinity. By these means, local lists have been prepared, and travellers who have made these points the especial object of their journeys, have furnished tolerably perfect data for the rest of our continent; thus we are in a condition to draw conclusions with much more certainty, and with a much greater probability of obtaining true results, than can be the case with the vegetation of more distant, and as yet partially explored regions. I shall therefore attempt to unfold the most interesting and important of these conclusions, and the speculations connected with them, and endeavour to overcome the difficulties arising from the technicalities involved, by incidental explanations and by the selection of facts and examples which are in some degree familiar to general readers.

The subject before us has then to be treated in its general bearings under two distinct points of view, namely the distribution as affected by physical causes, such as climate, and the distribution as a result of the spreading of different kinds by gradual migration, transport, or other processes from certain points or centres, where alone they have at first existed. In the consideration of these questions, it will be frequently necessary to allude to particular plants or groups of plants, selected as illustrative of the principles or as examples of the facts brought forward ; it is desirable therefore that a certain amount of acquaintance with these plants should be possessed by the reader before entering upon such inquiries. The limits of the present work preclude the possibility of entering very minutely into the characters of the European plants, and the only alternative that remains, is to seek for a number of generally known groups or species as representatives of the rest; this method, however, can only be made available to a certain extent, and it is impossible to avoid bringing into frequent use the botanical names of many characteristic species. Under these circumstances, the reader who is unacquainted with them can only be expected to follow the train of reasoning, and may pass over the names of plants unknown to him, as supplied merely to furnish more convincing evidence to those who enter more deeply into the technicalities of the subject.

It remains yet to speak of the influence of the hand of man. Important as the effects of climate are, and not to be combated beyond a certain point, yet is the struggle with the elements vigorously sustained, above all in the north of Europe; curious and striking as are the phænomena of the spontaneous migrations of plants, they sink into insignificance beside the operations of man's improving hand. Barren plains are forced to yield a crop of food-plants; bogs and marshes are drained and turned into arable lands; vast tracts wrested from the sea and brought under the dominion of the husbandman. The bare escarpments of the rocky banks of the rivers of Germany are terraced by the patient hand of industry, and converted into smiling vineyards. The whole face of the more populous countries is changed, and a view of the vegetation of Europe would be deficient in some of its most attractive and important features, without a sketch of the distribution and characters of the culti-vated plants.

Not without interest either is the history of the changes which have gone on in earlier epochs of the earth's ex-istence, before man trod upon its surface, and claimed dominion over the surrounding creation. Few and indi-stinct are the records on which this history depends, frag-mentary the traces of the vegetation of ancient times that remain to reward our investigations. Such as they are, they have been industriously studied, and, had we space, would claim a notice in the present volume. But in reference to this field of inquiry, we are compelled to limit our remarks to those more recent changes which have intimately affected our existing vegetation.

These are the points to be discussed in the following chapters; the arrangement of the matter is not the least difficult portion of the task; the plan which I have con-cluded to pursue is open to objections, and others have suggested themselves; but on the whole it appears the best that can be adopted in the treatment of such a miscellaneous subject, when a rigid scientific method would defeat the intention in view, and give an abstract and repelling aspect to what is designed for the instruction of general readers.

CHAPTER II.

GENERAL INFLUENCES ON THE DISTRIBUTION OF VEGETATION.

In those brisk, invigorating walks, on bright, clear winter-days, which form so prominent a part of the pleasures of a Christmas in the country, it is not rare to see the humble chickweed opening imperfect flowers, while every mossy bank displays the activity of its crowd of tiny inhabitants, lifting their elegant little urns on hair-like stalks, which, tender as they look, defy the frosts of winter. Yet at this time the delicately-nurtured favourites of our gardens demand the utmost care to shield them from the biting cold; or, if more hardy, have thrown off their showy summer garb, and either cower beneath the soil, or hide their soft buds, close wrapped in many a varied form of winter covering. Through more genial seasons, too, from when the snowdrop and the fragrant violet announce the coming warmth, through the full blaze of summer to the time when the autumnal tints and falling leaves bear witness of the waning year, we still must pile the constant fire to foster the rare and delicate strangers, captured from the crowd of beautiful, fantastic, or gigantic tribes that revel in full life, uncared for and in freedom, beneath the glowing heavens of the tropics.

Foremost of all the natural forces that limit or decree the general character of the vegetation of each region of our globe stands *heat*; this is unquestionably the principal constituent of that set of conditions we call climate. As such it was recognized by the first observers in the field of geographical botany; and if, as not unnaturally happened, even too exclusive a dominion was at first attributed to the mere proximity to the sun, still it remains the highest governing power. From the direction of the sun's course, the gradual approach of this source of heat to the northward on the ecliptic line in summer, and the retreat to the south of the equator in our winter, were there no other modifying causes, the comparative temperatures of different places upon the globe would be clearly marked by the parallels of latitude, and these lines would connect places of equal temperature, having similar summers and winters, and the same amount of daily warmth or cold. We should then find an uniform excess of heat at the tropics, and an uniform diminution toward the poles. We might mark the daily temperature at any place on one of these lines, and by taking the mean, or average of all the days of the year, obtain the *annual* temperature or amount of heat received, which would be the same at every place upon that line around the globe.

But applying to such a supposition the test of experiment, of observation, it was immediately found that places upon the same parallel by no means agree in their daily temperatures. The facts most familiar to horticulturists and travellers belied it, for it was known that plants will flourish perennially in the British Isles which are killed by the frosts of winter in places lying considerably to the south upon the continent ; thus the laurel, that bears our winters steadily in Ireland and the west of England, and is only affected by very severe frosts in our eastern counties, is killed by the winters of Berlin, equally fatal to the myrtle, the fuchsia, and a host of other shrubs which attain considerable age and size in the western portions of the British Isles.   Again, Canada, which lies south of Paris, has the climate of Drontheim in Norway ; while at New York, lying in the latitude of Naples, the flowers open simultaneously with those of Upsala in Sweden.   Moreover, it was known that those very countries suffering so severe a winter's cold, enjoy a summer's heat far exceeding ours, since the snow lies for months on parts of Germany which yet receive sufficient heat in summer to ripen the grape and Indian corn.

There exists then some other cause which powerfully affects the influence of the sun upon the earth, exercising an interfering power, to moderate and alter the direct influence of the solar rays. This is the configuration of the land and sea upon the globe, giving rise, by the different action of the sun's rays upon solid and liquid masses, to the numberless variations of condition in the atmosphere surrounding them, and by the irregularities of height of the earth's surface to the diverse conditions of exposure of this surface to all these atmospheric conditions.

Heated water is lighter than cold; therefore the surface of the ocean, warmed by the sun, remains comparatively unchanged, and being a bad conductor, the heat is but slowly transmitted to the lower parts; but the agitation of the waves tends to equalize the heat over larger surfaces, and evaporation contributes much to check the increase of the temperature; the mass of water is thus but slowly heated. At night, when the sun leaves the sea, the heat radiating from its surface renders this cooler, the cooled particles descend and give place to warmer particles rising from below, so that a continued source of warmth is exposed to the air in contact with the surface, while the bad conducting power of the aqueous vapour also retards the cooling; therefore neither in day nor night is there great excess of temperature.

With the land it is different. The solid surface absorbing readily, grows hotter and hotter beneath the sun's rays, since no motion takes place to equalize the distribution of the heat. At night the radiation or cooling is equally rapid, and the clearer the atmosphere, the less moisture present in it, the more freely does this go on, since the heated air rises uninterruptedly from the surface, to be replaced by the cold bright air from above, or cooler neighbouring regions. The atmosphere of the sea then has a more equable condition of heat than that of the land, and it should follow from this that the two extremes must lie on the open ocean and in the interior of large continents, and that we ought to find gradations between the conditions according to the intermediate position of points at which we make our observations. And we are in possession of abundant facts disclosing a striking contrast between the *continental* and *maritime* climates, and explaining some of the most remarkable contradictions to the first supposition, that the temperature would be proportionate to the distance from the sun, the primary source of heat.

In these investigations, however, the interfering causes prove to be so numerous, that the observation of existing temperatures is only a first step toward a much wider field of research; and the apparent anomalies require a very extended examination of the physical phænomena of the globe in order even to their partial elucidation. Some of them it will be necessary to examine before we can acquire any satisfactory understanding of the climate of Europe, but in the first place we will take the facts of the existing temperatures.

The phænomena of heat upon the globe have been greatly studied, and, aided by the developments of those principles laid down by Humboldt, which have occupied so much of the attention of meteorologists and naturalists of late years, we are able to obtain a very satisfactory view of the absolute and relative climates of different parts of Europe without much difficulty.

The observations of temperature obtained by the thermometer, and continued over long series of years, afford the data for characterizing the heat of the climate; but in their condition of long catalogues of facts, indicated by figures, they are but the crude, though indispensable materials for simpler and more general statements. The first process to be applied to them is that of obtaining averages or *means*; that is to say, of extracting from a set of observations of thermometric degrees an intermediate figure by which the combined or ultimate result of the whole is expressed. Thus to obtain the mean temperature of a day, say by hourly observation, the degrees marked at each hour are added together and divided by twenty-four, the quotient of which indicates the average amount received, since all the figures above that number are regarded as balancing or neutralizing

those below it. It is found, in practice, that a sufficiently accurate mean temperature for the day can be
obtained by an observation of the highest and the lowest
degrees indicated by the thermometer in the twenty-
four hours ; the intermediate number derived from these
two is near enough to the average of observations carried
on through the twenty-four hours.

The first attempt to establish a systematic comparison
between the temperatures of different climates was made
by Humboldt, who obtained for this purpose a number
of annual mean temperatures, deduced in a similar way
from daily observations carried on through a series of
years. In order to demonstrate the relations of these
clearly, he proceeded to lay down upon a map lines connecting all those places together which have the same
mean annual temperature ; these he called *isothermal
lines,* or lines of equal heat, and, as is seen by a glance at
the map, they at once declared themselves to be under
the most evident influence of the distribution of land and
water, and in a great degree independent of the parallels
of latitude in the northern hemisphere, an independence
however most marked as they approach the pole, and
becoming less and less observable toward the neighbourhood of the equator ; yet even there bearing witness of
the greater influence of the absorbing and radiating
power of solid matter, by the advance of the thermal
equator, or line of greatest annual heat on the globe,
almost wholly to the north side of the terrestrial equator.

The line indicating the points where the cold is so
intense that the ground is permanently frozen, leaves the
eastern coast of Labrador at about 54° N.L. and runs up
to a point more than 6° further north, near the south
coast of Greenland. Thence extending obliquely toward

the pole, skirting Iceland, it strikes the north coast of Lapland beyond 70° N.L., more than 16° nearer the pole than the point where it left the American continent. It then goes south again more than 15° as it gradually sweeps across Lapland and Siberia, whence it extends up north, reaching the western coast of N. America, there again to be deflected. According to the researches of Prof. Dove, the region of the greatest cold upon the earth, in January, is a tongue-shaped strip extending down from the pole to below Yakutsk in Siberia, within which the temperature has a mean of 44° below zero, Fahr., in that month.

It is unnecessary, however, to travel so far north to acquire striking instances of the irregularities of the curves of mean annual temperature. The isotherms of Canada pass through Iceland, across about the middle of Norway and Sweden, St. Petersburg and Kamtschatka. Those of New York through the north of Ireland and England, 12° further north, N. and Cent. Germany, and the Crimea. That which leaves the United States at about 36° N.L., crosses Southern Europe from the north of Spain to the Adriatic in a tolerably straight line, some 8° further north, and then falls south again, where the influence of the north-east polar current is more felt, in Greece and Turkey.

But if we depend upon these annual means for our ideas of the climates of different regions, we derive very erroneous impressions of their true nature. I have already spoken of the discrepancies between the coast and continental climates, arising from the *constancy* of the one and the *excessive* character of the other. And annual means are most directly open to this error ; for supposing that any place had a daily mean of 60° all the year

c

round, and another had a mean of 80° for the six summer months and of 40° for the six winter months, these
two numbers would give 60° as the annual mean in the
second case also, and the two places, so essentially and
importantly different, would lie in the same annual isotherm. Thus do we get a clue to the reason why the
culture of wheat and the vine is bounded on the north in
the European mainland almost by the 50th parallel of
N.L., while the yearly isotherm runs obliquely south-
east across the continent, and shrubs which brave our
winters never survive those of Germany.

It was at once seen, therefore, that no application
could be made of annual means to the elucidation of the
climatic influences upon vegetation ; they are too general, for not only does an excessive cold in winter destroy
perennial plants, but the life of annual vegetation is
wholly independent of the winter, so that their seeds be
but protected ; they spring up, flourish and decay during
the warm seasons of the year, and thus are almost completely under the dominion of the summer sun ; and a
country where the snow may lie for months upon the
soil, burying, but at the same time protecting the seeds
of annuals, may have a summer hot enough to ripen
fruits which never come to maturity in milder and more
equable regions.

The next point that was secured was the establishment
of lines indicating the means of the summer and winter
half-years, called by Humboldt *isotheres* and *isochymenes*,
or lines of equal heat and equal cold, which for conveni-
ence we may call summer and winter lines. These gave
results approaching much more closely to the facts of
vegetation ; but even here anomalies occurred which in-
duced the further prosecution of the inquiry, so that we

now possess, from the labours of Dove, maps indicating the mean temperatures for each month of the year. These will undoubtedly prove very valuable aids to new and more minute inquiry into the relations of vegetation in different countries. For Europe they may be regarded as pretty closely approximating to correctness, so far as they go; but those relating to the less-known portions of the globe being founded on observations made at distant points, and in many cases only for short periods, there is of course still much to be desired. The great value however of the monthly means in temperate climates is evident when we remember that the means taken for longer periods usually contain degrees below the freezing-point, which are counted as neutralizing so many above that point at other times. Now when vegetation is merely brought to a standstill by cold, and not destroyed, the above method would often lead to a false result in calculating how much *useful* heat a plant will receive, since all temperatures a little above the freezing-point are useful, while all below are not *injurious*, but merely *indifferent* in such a case, and should be left out of view.

In the accompanying map, the isotherms of January and July only have been laid down, since to have given the monthly lines on one map would have been to have rendered it a maze of inextricable confusion; and these are sufficient to indicate the most striking features of the question, as the other months only form intermediate steps between. There have also been inserted what Prof. Dove calls the *thermal normals* for January and July, which have the following signification. By remarking the temperature at a number of places upon the same parallel of latitude, we may obtain an average of the heat received by the whole of them, and this will be the mean temperature of

c 2

that parallel; places which actually have exactly that temperature have the *normal* or proper temperature of that parallel; those where it is lower are colder than they should be, those where it is higher are too warm. Then, if we regard places which are too warm in winter and too cold in summer, that is, are *equable*, as possessing a sea climate, and those where the summer is too hot and the winter too cold, where the changes are *excessive*, as having a continental climate, the *thermal normal* lines, connecting all those places which have their proper temperature, will be the boundary-lines between regions exposed to these two kinds of climates.

The more particular description of the map is reserved for the next chapter, containing an application of the general views exposed in the present, to the continent of Europe alone, in greater detail. Let us turn now to a consideration of the chief modifying causes that affect the heat of climates.

In his Voyage in the Levant, Tournefort relates that he was struck with the alterations in the characters of the vegetation as he ascended Mount Ararat; at the foot he passed through the plants of Asia Minor; half-way up he met with those of France, and at the summit he recognized the members of the Lapland flora. Linnæus carried out this idea somewhat further, and the observations of all subsequent travellers have only served to confirm and extend it. Step by step as the land rises in any mountain region, the vegetation assumes, more and more, a polar character; and in the mountains of the tropics, a succession of stages have been distinguished, corresponding in the general peculiarities of the plants which clothe them, to tracts extending horizontally, in succession, on the sea-level, from the base of these mountains to the frozen regions within the arctic and antarctic circles. Increase of elevation is accompanied by an alteration of climate, bringing with it a set of conditions analogous to those prevailing at certain distances further from the sun. Ascending the Peak of Teneriffe a series of regions are traversed, one above another, displaying with the approach to the summit a continually closer approximation in character to the polar regions, till the traveller who left the palm, the cactus, and the thousand varied forms of tropical vegetation at the foot, finds himself at last among the stunted shrubs and scaly lichens, the borderers who hold the outposts on the limits of the eternal snow.

This difference of vegetation is the result of a difference of climate, in great part of a diminution of heat in elevated regions, to be explained on the same principles as the irregularities of horizontal distribution. The atmosphere enveloping the globe diminishes in density with its distance from the earth's surface. Air is much thinner and much lighter at great altitudes. Now air is at all times a bad conductor of heat, and the tendency to become heated is lessened as it becomes thinner; thus the air is much less heated by the sun's rays at high elevations. But radiation goes on more rapidly through thin media; so that, although the soil of high mountains becomes strongly heated by the sun's rays, it cools as rapidly; moreover, evaporation of moisture, a very energetic cooling process, also goes on more freely as the pressure of the atmosphere is diminished, and contributes greatly to lower the temperature of the ground. Added to which, the radiation from the earth's surface, the giving back during the sun's absence, of the heat received in the sunshine, is an important source of the heat of the cli-

mate, and the mountain summits, projecting from the
general surface, are removed, in proportion to their height,
further from this influence; their own bulk, so freely
exposed as they are to all the cooling influences of radiation, evaporation, and currents of air, being far from sufficient to enable them to accumulate such a store, as it may
be called, of heat to be gradually parted with, as the
level plains below. This is strongly exemplified by the
much greater diminution of temperature in proportion to
the height, felt on steep mountain-slopes, than on elevated
plains, and the much greater equality of alpine cold climates than those of level regions further north.

The slopes of mountains facing to the equator will
receive more heat from the sun than those looking to the
poles. This is a strongly-marked instance of the effect
of difference of *exposure,* an element in climate of which
every gardener even is reminded by the different value
of the different *aspects* of the walls to which he nails his
fruit-trees. But in considering the phænomena included
under this head, it becomes necessary to take into consideration the other effects of the movements of the earth
and the independent motions of its atmosphere, and indeed of the mobile waters of its seas. The nature of
prevailing winds is very important in determining the
characters of climates, not from the mere influence of
wind as such, but from the varying conditions of warmth
and moisture which are produced by the movement of
currents of air to and from differently constituted regions.
Wind blowing over the land from a widely-extended surface of the ocean comes loaded with moisture, and if it
meet with a colder body of air, the watery vapour is condensed and falls as rain, the heat at the same time being
imparted to the colder air and raising the general tem-

perature of the atmosphere. Air passing over a broad continent comes desiccated to the other side, robbed of its moisture, whether the land be cold or hot; for in the first instance the watery vapour will have been precipitated as rain, and in the second it will have been greedily absorbed by the dry atmosphere, possessed of far less than a sufficient share of humidity. Thus the moist south-west winds, which bring rain and damp warm weather to Western Europe, reach the eastern parts deprived of all excess of moisture; and the winds of Northern Africa bring up cloud after cloud to the borders of the parched deserts, there to be dissipated by the intense heat, and reach the coast as hot arid breezes, or rising into the great equatorial current, fall again farther north on Italy as the sirocco.

When speaking just now of the distribution of heat, I stated that were it not for modifying causes, the temperatures would be regularly proportionate to the latitudes. With the winds again there exist certain primary influences, qualified and interrupted in their action by secondary causes. Were the temperatures equal in equal latitudes, and the earth's surface uniform, the winds, contrary as the statement seems to our common notions in a northern climate, would be regular and steady,—would be no more the " idle wind " that " bloweth where it listeth." And science, which may not let this " idle wind" "pass by" without challenge and inquiry, has shown that as things are actually constituted, order and law prevail, although complex and intricate in their operations, even over the wind, which in every age has passed for the type of all that is fickle and uncertain.

The principal cause of the movement of the atmosphere is the disturbance of equilibrium by the unequal amount of heat received in different regions. The sun heats the atmosphere more strongly in the equatorial region than in any other place ; this heated air, becoming lighter, rises, and the cooler air from the polar zones rushes in to fill up the partial vacuum. Thus a circulation is established, since the heated air which rises in the tropics becomes cooled above, and then falls again towards the earth near to the poles. But the rotation of the earth, combined with its shape, affect the direction of the currents thus set up, which otherwise would be direct north and south winds. The air which revolves with the earth near the poles travels with little velocity, since the distance to be passed through in the twenty-four hours is comparatively small ; but when this air begins to rush toward the equator, drawn in from the poles by the rarefaction taking place in the tropical zone, it is only by degrees that it acquires the additional speed necessary to keep pace with the rapid motion of the earth's surface at the greater diameters, where a complete circuit is still performed in the same time. The earth thus slips away, as it were, from beneath the wind, and as in looking at objects from a carriage-window we may fancy they are running past us, so the lagging atmosphere, left behind in the revolution of the earth, seems to rush over its surface and becomes a wind ; the direction of this, contrary to that of the earth's motion, is from east to west, which, combined with the original direction from the poles to the equator, gives an intermediate direction, a north-east wind in the north hemisphere, and a south-east wind in the southern hemisphere, the *trade-winds*, which, meeting together in the tropics, produce a general current of air from east to west. In the very equatorial region indeed, for about 2° on either side, there is a kind of equilibrium produced by the opposition of the two currents and the general tendency of air to rise upward; hence those fearful *calms*, which voyagers have described in many melancholy tales of suffering. But this region is also visited by violent and irregular storms, the unstable equilibrium being disturbed by very slight causes. Hence this has been called the region of calms and storms.

The heated air, forming the equatorial currents going north and south in the higher regions of the atmosphere, passes over the top of the trade-winds in return currents, resulting from the same causes acting in exactly the opposite way. Their speed being greater than that of the earth in the higher latitudes, they outstrip it in the same proportion as the polar currents fall behind, and thus produce a south-west wind above the northern north-east trade-wind, and a north-west above the southern south-east. It is not necessary to speak here of the causes which modify the regularity of the trade-winds all over the globe; it will suffice to direct attention to those ascertained to exist in the Atlantic Ocean.

The first point to be noticed is, that the zone of calms and storms lies altogether on the north side of the equator in the Atlantic, the space occupied by it being, as a mean, the interval between 3° and 8° of N. Lat.; but, varying with the seasons, it lies from 3° to 13° N. Lat. in August, and from 1° to 6° in February. This northern deviation of the zone is attributed by Humboldt to the conformation of the continent of South America; the direction of its coasts appearing to favour the advance of the south-east trade-wind and of the heated water of the great equatorial current further toward the north, thus heating the sea of the Antilles; while the high mountains which rise in the region of the equator exercise a cooling action upon the atmosphere, and draw a current of warmer air from the south. At all events the line marking the greatest mean temperature of the globe does pass through the south of the sea of the Antilles. This general higher temperature of the northern hemisphere is however in great part a result of the different distribution of land and water on the globe. In the northern hemisphere the sun acts upon a preponderating mass of land, which undergoes no alteration, and the heat therefore acts in raising the temperature; but in the southern hemisphere, where the liquid surface is so much more extensive, far more of the heat is occupied in evaporating water, in which process it becomes latent, not contributing to raise the general temperature; moreover, as a greater quantity of rain, of the water thus evaporated, falls in the northern hemisphere, this region gets the benefit of the heat which had become latent in the south, and is now set free in the condensation of the vapour.

With regard to the return currents in the upper part of the atmosphere, we may mention certain facts which have been recorded, strongly warranting the conclusion of their existence. Most travellers have found a west wind blowing at the summit of the Peak of Teneriffe, while the north-east trade-wind was blowing at the sea-level. The ashes of volcanos, thrown to a vast height during eruptions, have been carried in directions only to be accounted for by the existence of such currents; as when those of the island of St. Vincent have fallen in Barbadoes, to the east of St. Vincent; and when, in 1835, the volcano of Cosiguna, in Guatemala, threw into the air an enormous quantity of cinders, they were seen a few days after covering the streets of Kingston, in Jamaica, situated to the north-east of Guatemala.

Leaving the tropical regions those of the *constant* and *periodical winds*, and advancing to the north, we find as we proceed into a higher latitude a gradual diminution of the regularity, until the *temperate zones* acquire the character, whence they have also been denominated the regions of *variable winds*, but might properly be called the battle-ground of the winds, for here there is a constant conflict between the polar cold current and the warm equatorial current, which, cooled in the higher strata of the atmosphere, now falls nearer to the earth's surface. No rule has yet been established as regulating the frequent and sudden changes of direction of the wind in these regions ; it is only known from observation of the prevailing winds, that in the northern hemisphere the west and south-west and the east and south-east winds blow more frequently than those from other points of the compass ; and that on the north Atlantic ocean the westerly gales prevail greatly over the easterly, so that the sailing passages from America to Europe, and *vice versâ*, present, on an average, great disparity in the time they occupy.

The shape of the eastern coast of North America favours the progress of the south-west wind up the great valley of the Atlantic ; and this upper equatorial current falling to the surface about 30° N. Lat., or in summer still further north, is carried forward, greatly assisted by the Gulf-stream, into very high latitudes on the western coast of Europe, and loaded as it is with aqueous vapour, is one of the most important agents in determining the climate of this continent from Norway down to the Straits of

Gibraltar. On the other hand, the north-east polar currents, coming from cold regions, are found to take their way mainly over the continents, and extend down a great distance to the southward on the surface of the earth, unless checked in their course by barriers in the shape of high mountain-chains. The temperatures of the different parts of Europe are strongly influenced by these circumstances, as may be seen by examination of the directions of the isothermal lines. The line of the freezing-point in January runs from Baltimore, south of New York, obliquely upwards over the bank of Newfoundland across the Atlantic to the south of Scotland, and on for some distance beyond the north polar circle; then, arrested by the influence of the polar current, it is directed suddenly to the south, skirting the western coast of Norway, crossing the North Sea to enter the main land at Amsterdam. From thence it extends obliquely southward and eastward by Vienna to the plains of the Danube and the Black Sea, following across Central and South-eastern Europe the direction of the mountain-chains, which oppose an obstacle to the fierce northern blasts. This line coincides almost exactly with one drawn through the Böhmer Wald, the Thuringer Wald, and the Bakony Wald down to the Alps, and from thence along the Croatian Alps and the mountains of Bosnia and Bulgaria (the Balkan) to the Black Sea. (See Map).

The next line also, indicating a mean temperature in January of 23° Fahr., extends from Halifax in Nova Scotia up across Newfoundland to a point far north of Iceland and even of the North Cape, but turning to the south before it reaches the coast of Europe, it passes down clear of the coast of Norway above Drontheim, then entering the land it descends to Christiania inside the mountains of Langefeld, then suddenly crosses to Upsala, running almost due east across the Baltic to the island of Dago. Next driven south by the polar current, through the island of Esel, it enters Europe by the Gulf of Livonia and extends southward as far as Warsaw; then again directed apparently by the Riesen-Gebirge and the Carpathians, it extends eastward so as to pass north of the high ground of S. Russia lying on the north of the Black Sea. The equalizing influences of marine atmosphere, and the excessive nature of continental climates, arising from difference of condition and radiation, of course contribute largely to cause these irregularities.

A comparison of these January isotherms with those of July demonstrates very strikingly the increased influence of the N.E. polar current of air upon Europe during the time the sun is shining on the south side of the equator. In June, the S.W. equatorial current causes the lines not only to run obliquely to the N.E. across the Atlantic, tolerably parallel with those of January, as far as the seas bordering the coast of Europe; but now the lines, no longer directed south in that region, run still further north in the same direction over a larger proportion of Europe. The different influence, however, of land and water, when exposed to the more intense rays of heat, is very manifest in this comparison, and the lines extending across both ocean and land are nearer to the parallels of latitude in summer than in winter, since the equalizing action of the water moderates the temperature of the Atlantic; while the absorption of heat by the land, which by radiating so freely contributed to increase the cold in winter, now raises the temperature of the climate sufficiently to neutralize the effects of polar currents and to exalt the heat above its normal degree, that is, above the average temperature of places lying in that latitude, in that month.

But the conflict between these two great currents of the atmosphere, the war between these two giant forces, is continually waged, with varying success on either side, on the battle-ground of our temperate climates, giving to it an ever-changing character. The south-west gale, laden with grateful moisture, loses more and more of the south impulse as it advances toward the north and changes to a west wind; soon the north-east polar current charges down upon it, forcing it to drop the watery burden gathered from the ocean region, and probably overcoming it, converts it to a north-west wind which is wet and stormy. By and by the polar current gains another point, and its frozen breath brings cold (in the west often rain also from further condensation), the atmosphere becoming brighter and drier, but more piercingly cold, as the north-east wind, the lower trade-wind, the combined force of the polar and the circulating currents, gradually acquires the complete mastery. But only for a time does it hold the vantage-ground; waxing weaker and weaker it gives way, turns and becomes east, then south-east, south, losing point by point till again the south-west upper trade-wind is lord of the field. In these several changes are experienced the various conditions of heat, cold, drought, or humidity, resulting from the character of surfaces over which the winds have blown; the dry easterly wind is succeeded in north-eastern Europe by a balmy wind warmed by the "sunny south;" but in Italy this south wind is hot and dry from the proximity to the African deserts, while for us its heat is tempered by the sea, and in its struggle with the cold east wind, rain is frequently precipitated by condensation.

So much is known of the course of these uncertain winds, that there is a rotation, as it has been called by its discoverer Prof. Dove ; that the changes are always in one certain direction, from east through south to west, and from that through north to east again. But the rotation, the whole series of changes, may be run through in a few hours, or the.wind may blow in one direction for days and even weeks, and we know not the why and wherefore. The prevalence of a north-east wind dries up the melting snows in spring; but if it last too long, the excessive drought endangers and retards the rising crops. A south-west wind, bringing April showers, is in the highest degree favourable to vegetation at that season ; but if it blow through the months when the corn is coming into ear, the humidity of the air will prevent the maturation of the crops and spoil the harvest. Yet this same south-west wind, which will cause a general failure of the crops through Western Europe, losing its excessive moisture as it sweeps across the continent, may highly favour the husbandman of Southern Russia. Such was the case in 1816 and 1817. And when and where such conditions will occur in the temperate climate, is a question that science cannot answer, but, until some far greater advance in meteorological knowledge has been made, must unwillingly leave to the idle or interested fancies of the " weather prophets."

In considering the subject of currents, those of the ocean have also to be taken into account, since they greatly influence the distribution of heat and moisture. Of the phænomena of tides, it is unnecessary to speak here ; they are vast and universal movements, depending on the attraction of the heavenly bodies. But besides these, there exist determinate currents in particular directions, depending partly upon differences of pressure of the atmosphere at different parts of the globe, but more particularly upon the differences of temperature of the polar and equatorial regions. As is the case with the atmosphere, the ocean is subject to a process of circulation, arising from the action of tropical heat upon its upper strata. Like the air, the water beneath the equator is warmed and rendered lighter. The cold and heavy waters of the polar regions constantly strive towards this rarefied portion and displace it, receiving in their course toward the equator a similar deflection to the west to that which affects the lower trade-wind or polar air-current, and from the same cause, the increasing diameter of the globe as they travel onward, till in the tropics the two diverted polar currents meet side by side and cause a great current from east to west, the *grand equatorial current* as it is called. But this equatorial current is interrupted by the continents which stretch across its course, and more affected, since it is wholly arrested, than the trade-winds, by the shape of the coasts on which it impinges ; and this is especially seen when we examine the results of the investigations that have been made upon the Atlantic Ocean, with which we are most immediately concerned.

The equatorial current of the South Atlantic starts from the western coast of Africa, and flows at the rate of from two to three miles an hour towards America : striking upon Cape Rocca, it divides into two branches, one flowing south ; the other, the more considerable, sweeping along the coast of Guiana to the W.N.W., enters the Caribbean Sea and thence the Gulf of Mexico, around which it flows, past the mouths of the Mississippi, and emerges at the narrow opening between the extremity of Florida and Cuba, under the name of the Gulf Stream.

The Caribbean Sea and the Gulf of Mexico thus form an immense cauldron as it were, in which the water is strongly heated, and whence, constantly in motion, it is poured out over the North Atlantic. The vast body of warm water of the Gulf Stream rushing out parallel to the shores of Florida, but at some distance from them, and moving at a velocity varying from two to five miles an hour in different seasons, passes beyond Cape Hatteras as a narrow, deep and rapid stream ; for it has been shown by careful thermometric soundings, that it possesses the same superiority of temperature above the surrounding seas, at a depth of 3000 feet below the surface, at this point, while its small breadth is ascertained by the existence of this same temperature only over a narrow belt of water. Here arriving in the open ocean it meets the cold polar currents coming from the north and the sand-banks which extend along the east coast of America as far as Newfoundland. It is thus diverted to the east, and by virtue of its less density rises and flows over the colder waters going in the opposite direction, so that, spreading out, it goes on in a broader but slower course to the Azores, whence it runs south to recommence on the coast of Africa its never-ceasing circulation. To compare great things with small, if the southern hemisphere resembles a distilling apparatus, raising the vapours from the ocean, to be condensed in pure dews and rains on the continents of the northern hemisphere, we may again liken this rotation of the currents of water to the boiler and hot water tubes with which we heat our houses and conservatories, where the water rendered lighter by the heat rises from the boiler, carrying the heat to distant points, along the

D

pipes, and then gradually cooling and becoming more dense falls into the return-pipes to the boiler to receive fresh heat, and then to start again on its unceasing circulation.

The warm water of the Gulf Stream travels still further than the points just mentioned: under the influence of a south-west wind it is carried to the coasts of the north of Europe, and contributes to elevate the temperature of the climate ; and even not unfrequently leaves a more remarkable evidence of its visit there by the deposit of seeds or fruits transported from the West Indies. The return polar current flows down on the shores of North America between the Gulf Stream and the land, and again outside it ; and not only there but under it, as is demonstrated by the movements of vast icebergs, the lower parts of which are sunk deep in the under-current and thus are carried onward and enabled to make headway, through and against the weaker portions of the Gulf Stream. Along the whole western coast of Europe the influence of the upper warm equatorial stream battles upon the surface against the polar current continually flowing southward from the Arctic Sea.

The conditions of climate arising from *rains* are very different in the tropics and the temperate regions, as indeed follows from the different characters of the winds, upon which they chiefly depend, as has been shown already. In the tropical regions, with the constant winds, there occur periodical rains. While the trade-wind blows steadily, the sky is serene, bright and dry. As the sun approaches the zenith, the weather becomes more uncertain, for the region then falls under the influence of the conflict between the opposing currents. Storms then occur and vast quantities of rain fall, lasting for weeks, the "rainy seasons," in which vegetation is forced into the highest activity by the combined effects of heat and moisture. As the sun retreats the sky clears and the trade-wind resumes its sway, and all is dry again until the commencement of the next rainy season.

In the temperate regions, the abode of the "conflicting winds," little trace of periodicity can be detected. The general character of the configuration of the surface of the earth, the distribution of land and water, the peculiarities of prevailing currents of air, or water in contiguous seas, so greatly influence the distribution of rain, that it is very difficult to extricate any clear and accurate general statements. It is known that on the south-western coast of Europe, at Lisbon for instance, heavy rain-storms occur as "winter rains," arising probably from the meeting of the upper trade-wind with the north-eastern gales. Italy and certain portions of the Mediterranean have spring and autumn rains, which Dove attributes to the passage of the south-west trade-wind before and after the summer solstice. He also states that the frequent rains about the summer solstice in Germany—for in central Germany July is regarded as a wet month—are caused by the S.W. trade-wind, which then reaches its highest point in these latitudes, at the time when the sun is nearest to the north.

Yet the most striking differences in fall of rain in different parts of Europe appear to depend more particularly on the elevations of its surface. With regard to the mountain ranges, the elevated chains of Norway, the Alps, and the Apennines, have been shown by observation to exert great influence.

The south-west wind, coming loaded with moisture from the Atlantic, meets with the great ridge running

down the western coast of Scandinavia, and, rushing up
the cooler mountain-sides, the vapours are almost entirely condensed, causing a climate of almost continual
rain or drizzling mist. At Bergen the sun is rarely seen
unclouded, and 82 inches of rain fall there,—a large
amount, which is nearly equalled all along this coast; and
since this great condensation sets free much latent heat,
the atmosphere is proportionately warmed, a soft and
steady, although damp climate being the result.

To the other side of the mountains and to Sweden the
same wind comes deprived of its moisture and brings
bright cold weather, and a climate where only 21 inches
of water are received.

The warm and moist winds from the south-west, blowing from the Atlantic and the Mediterranean over northern Italy, are arrested and gathered to a focus by the
curved chain of the Alps. There the snowy summits lift
a wall impassable to the aqueous vapours, which are
condensed in rains and flow down the southern slopes to
fertilize the plains of Lombardy. Only 36 inches of
water fall on these plains; but at the very foot of the Alps
an average of 58 inches is received. In the angle at the
north-east corner, at Tolmezzo, in the valley of the Tagli-
amento, the vapours accumulate to such an extent that
their condensation gives an average of 90 inches of rain
every year, as shown by observations carried on for
twenty-two years. On the north side of the Alps only
35 inches are received at the foot.

The chains of the Apennines furnish further examples;
the arch formed to the north of the Gulf of Genoa bars
the passage of the vapours from the sea, and 64 inches
of rain fall at the southern foot, while only 26 are received on the northern side, in the plains extending to

the Po. Farther south, where the same chain stretches more nearly north and south, 35 inches of rain are received on the west side, and only 27 on the east.

The influence of plateaux or table-lands is illustrated by the conditions of Spain. There the central plains rise from 2000 to 2500 feet above the level of the sea, and while the south-west coast of Portugal, at Lisbon, receives 27 inches of rain, the border of the table-land has only 11 inches; and Madrid, in the centre of the arid plains of Castile, only 10 inches, the smallest amount of rain that falls in any part of Europe. Yet at the same time Spain also affords the maximum amount, for it has been stated that at Coimbra, in the valley of Mondego, at the western foot of the Sierra d'Estrelles, which projects as a promontory far towards the sea-coast, measurements taken in 1816 and 1817 gave 225 inches; and although this account has been supposed to have arisen from some error, Schouw reducing it to 135·7, and Kämtz to 119 inches, this last number gives a difference of more than 100 inches from Madrid, so little removed geographically from Coimbra.

There is still another force which must not be neglected, mysterious and inexplicable as many of its effects at present are. The commonest observation of the life of plants tells us that *light* exercises a most important power, independently of heat, over all the chemical and even of the physiological phænomena of vegetation. With regard to the last it may suffice here to mention the opening of flowers in obedience to the sun's rays, the closing up at night, and the occurrence of these operations even at particular hours of the day. We have no explanation of these facts, but they are probably attributable to the influence of light.

The chemical changes going on in the interior of plants are almost wholly under the dominion of light. In darkness plants become weak, watery and colourless, and under most circumstances soon die. The green colour, indicating the accumulation of elaborated nutriment, seems to be directly proportionate to the intensity of light; and although by artificial heat we are enabled to ripen in our stoves the fruits of the tropics, the deficiency of light is strongly manifested by their soft and succulent condition. Our hot-house pine-apples differ in firmness and solidity of texture from those of the West Indies, as the blanched endive or celery does from the tough and acrid plants of the same species growing wild. All travellers mention a brilliancy of colour in Alpine plants, which has been attributed to the more active influence of the sun's light at high elevations; but whether these statements are really accurate, or the colours depend on the peculiar characters of Alpine species, preservable when they are removed to other stations, there is no doubt that the clearness of the atmosphere, as favouring the direct influence of the sun's luminous rays, must be highly important in determining the properties of plants. The hemp is said to develope little or none of its active narcotic principle beneath the sun of Europe; and the increase of resinous and aromatic shrubs in the dry and bright warm regions is probably as much dependent on the light as on the increased temperature, since those principles appear to be elaborated almost wholly through the agency of the chemical rays of the sun. From beneath the sun of "Araby the blest," and the cloudless, arid regions around, come "myrrh, frankincense and gums." The scorched plateaux of Spain abound in the aromatic Labiate plants—the lavenders, the sages, and their allies; the "wild thyme" with us marks the sunny bank ; and the Old World, with its much wider expanse of dry tropical regions, is a far more fertile source of plants possessing aromatic properties than the tropics of the New World, where the greater moisture, in most instances, favours the luxuriance of vegetation, but by the increased quantity of vapour it pours into the atmosphere seems to arrest to some extent the influence of light.

Of the direct effects of *electricity,* and of *atmospheric pressure* upon plants nothing is clearly known ; it is probable that they are chiefly influential through the medium of the variations of humidity and temperature to which they give rise, and their phænomena are not sufficiently general to require a separate mention here.

It remains now only to make some few remarks respecting the general influence of soil in connection with climate. An elaborate account of this subject is altogether unnecessary for our purpose.

The well-known fact of the exhaustion of soils by repeated cropping with the same plant, and the necessity of fallows, or of an alternation of crops, would suffice to draw attention to the importance of the nature of the soil in determining the character of the vegetation upon it. The almost exclusive growth of certain plants in a natural condition on soils characterized by some peculiar mineral character, such as that of the maritime plants peculiar to sea-shores or the vicinity of salt-springs, and the less definitely marked preference of many inland plants for particular rocks, to speak geologically, are further illustrations of the same principle. Plants require for their support not only carbonaceous and nitrogenous substances, in addition to the water with which these unite to compose the great mass of vegetable structure, but also certain proportions of earthy ingredients, and the results of chemical analysis have proved that different substances of this kind are actually required in different proportions by different species or tribes of plants.

This might at first appear to be a sufficient explanation of the phænomena above adverted to, and the chemical theory has been strongly advocated by many botanists. But while admitting the force of the evidence as regards cultivated plants, most observers are now inclined to attribute a far inferior influence to the chemical constitution of soils in the natural distribution of plants; for the instances of exhaustion of soil by culture are produced by *excessive* artificial cropping, such as is never seen in nature, where plants almost universally grow more intermingled, and often intermit their growth so as to afford conditions analogous to rotations of crops and fallows; and more widely extended and more accurate investigation has shown that the preference of certain plants for rocks, having a peculiar chemical character, is rather apparent than real; in fact, that the influence is rather conveyed through the conditions of humidity produced by the physical character of soils and subjacent rocks.

As on a large scale with the climate, so on a more restricted one with the soil, the degree of moisture appears to be the most important element next to the heat; and in considering these influences, it is at once evident that the characters of the soil and climate must react one upon another to produce conditions altogether different when they are opposed from those resulting from a co-operation of their effects.

Thus a light sandy soil, which will be loose and shifting, scarcely affording hold or nourishment to a single plant under a hot and dry atmosphere, may under a humid climate be capable of sustaining a tolerably abundant vegetation, deriving its support chiefly from the moisture continually supplied to the absorbent soil by the atmosphere.

The influence of climate is also manifest, though perhaps in a less striking manner, upon tenacious and heavy soils : under a damp cold climate, their vegetation will be coarse, consisting of plants that require little heat for their maturation ; under a warm moist atmosphere their vegetation will be highly luxuriant ; the moisture which they retain abundantly combines with the usually plentiful supply of soluble mineral matters, affording the necessary support for a vast quantity of plants upon a given surface, under the vivifying influence of the heat ; while a heavy soil, under a hot dry climate, will usually still retain within itself enough of the moisture supplied to it from time to time to sustain an active vegetation, usually of a character most valuable to man, in which heat and dryness are required for the maturation of the fruit or seeds, but which cannot find support in a similar climate upon a soil readily yielding up its moisture to the atmosphere.

It is evident, if we admit these facts, that the difference of soils must diminish more and more as they are exposed to a greater humidity of climate, and that the greatest contrast between their vegetable products will be found in dry climates and on plains, while the difference will gradually become less marked in advancing into a humid or *coast* climate, or ascending to elevated points, where the supply of moisture by condensation of aqueous vapours becomes more abundant in proportion to the height, especially on particular exposures.

The foregoing brief sketch shows how vast and varied is the field of inquiry into which the study of physical conditions leads. I have but touched upon a few points of the earth's surface in the illustrations, but these will serve to unfold the closeness of the bond that here unites all the natural forces together, and how influential they are upon one another. The more powerful agents enforce their general laws, but every little local action asserts its qualifying voice; and we see that all these irregularities and uncertainties, as we in our ignorance call them, and complain of, are necessary and important parts of a great whole,—are but isolated features of a comprehensive plan, in accordance with which all work in concert to bring about that *change* absolutely indispensable to the existence of animal or vegetable life upon the earth's surface, and that *variety* of conditions by which is ensured a fitting abode for each kind of its multifarious and diversified inhabitants.

CHAPTER III.

SPECIAL INFLUENCES ON THE DISTRIBUTION OF PLANTS.

In the foregoing chapter we have only taken into account the circumstances which *allow* of particular conditions of vegetation in different regions; there remains another, and not less important class of phænomena to be examined, namely the circumstances which *cause* the peculiar vegetation of particular places.   In the former case, the objects of investigation were the general laws to which all plants lie subject : now we have to inquire into the history of the plants themselves, as it has been enacted under the influence of those laws ; and here we find a new element in the question, the individual character and peculiarity possessed by each particular plant, which makes itself visible in the absolute condition of distribution of each species.

Reflecting merely on the principles that have been developed in the preceding pages, it might be imagined that every plant would be found growing on all the portions of the earth's surface which afford it the proper physical conditions.   It might do so, certainly, and in the instances of some of the more important cultivated plants, such is almost completely the case ; but this is owing to the hand of man, who has carried with him to each land he has invaded, those *tamed* subjects of the vegetable world on which he chiefly depends for his sustenance.

But the natural condition of vegetation offers a much more complicated picture.   The "carpet of nature" has been woven with a far greater care for the variety of its patterns, and to the diversity arising from external causes is added an additional and much more curious diversity, manifesting itself in the occurrence of plants different in form, specifically different, but of similar habits, in like conditions on different parts of the globe ; plants which *represent* one another, as parts of the general scheme of creation, but differ decidedly in their own individual structure.

If we examine, for example, two spots agreeing almost

exactly in climate, one in the temperate region of Europe and another in the temperate part of N. America, we shall find that the greater portion of the indigenous plants differ in the two localities; perhaps they will not have even one plant in common. Of some 3000 species described as inhabiting the United States, not 400 are at the same time natives of Europe. The oaks and pines of Europe are represented by other species of the same genera in N. America, and so with a large majority of the other plants. Even in Europe itself some of the arctic species are represented by different species in the arctic climates of the Alpine regions of the South, and while the *Rhododendron lapponum* is characteristic of Lapland, the *Rhododendron ferrugineum* and *hirsutum* take its place on the Alps of Germany and Switzerland. Were we to glance over a wider field, the instances of this phænomenon would be far more striking, and we should find that there exists a representation not only by species, but by groups of plants, such as genera and even families; as for example the accumulation of the plants of the true Heath family or Ericaceæ at the Cape of Good Hope, and the total absence of these in the presence of the Epacridaceæ, a closely allied family, under similar conditions in New Holland. The predominance of Cactaceous plants in Central America, and of the spiny, misshapen Euphorbias in tropical Africa, afford another very curious illustration.

The examination of vegetation in general leads therefore to a new mode of considering their distribution, and we proceed to divide the earth's surface into vegetable regions, not characterized by their climatal conditions, as the tropical, temperate, arctic, &c. are, but by the predominance of particular forms of vegetation; into regions in which, while they contain a vast variety of plants, those belonging to certain groups appear in the greatest abundance and development, and thus form the most characteristic feature of the vegetation as a whole. In this way the world has been mapped out into provinces, denominated according to the families which appear in the greatest force there; and while the knowledge of the habits of these families gives us a key to the climatal conditions of each region, and thus an indication of the general aspect of the vegetation of each, the fact of the predominance remains as a peculiarity,—as a residual phænomenon to be accounted for by some other cause.

How are these peculiarities explained? Why do we find a particular plant growing only on some particular portion of the earth's surface, while the conditions fitted to maintain it exist in so many other localities?

These are questions which cannot be answered by a demonstrated history of the circumstances, and the only path that remains open is that of inquiring as to what hypothesis will best explain the facts which we have existing before us. We must seek out some view of the original plan upon which the creation of plants was conducted, guided in our search by the actual conditions of vegetation and our experience of the influence of their own peculiar qualities, and that of external agencies in modifying the conditions with the lapse of time.

By means of the floras, as they are called, that is, lists of the plants of particular regions, which the labours of botanical inquirers have now so multiplied, we are enabled to form a tolerably exact conception of the actual distribution of species in Europe, and an approximate view of that of many other countries. The comparison of these demonstrates the existence of *areas* of distribution for particular species, that is, of definite tracts of land upon which alone these species respectively grow. These areas also often exhibit a centre or focus in which the species occurs most abundantly. To plants which are characterized by this condition of distribution,—and the great majority of known plants come into this category,—the name of *endemic* or local plants is applied, and on these is based the hypothesis which has now to be explained.

It is believed by most inquirers in this field, that each plant was created at a particular point, on that spot which now forms the focus of its distribution, and that from this, its specific centre, its descendants have spread out in different directions, by the combined agency of its prolification and the diffusing influences of external forces under the limitation of climatal conditions.

Some naturalists contend that the original creation must have consisted of a number of individuals; but it seems more in accordance with the simplicity of nature, to suppose that a single parent or pair of parents alone was produced for each species. A more weighty difference of opinion arises from a series of facts, exceptional to those above enounced of the *endemic* or local distribution of plants, namely the instances, which are numerous, of what is called a *sporadic* or *universal* distribution of a species, as, for instance, of the sea sedge (*Scirpus maritimus*), which occurs in similar situations nearly all over the world; as is the case almost to an equal extent with *Samolus Valerandi*, and in a less degree with many others. Such very wide distribution of certain plants has led to the opinion that species were created in different places, at different centres, either simultaneously or at different epochs, and that thus different families or stocks of the same species coexist and share in populating the area inhabited by the species. The instances, however, on which this hypothesis rests, and the almost unlimited powers of diffusion that exist, together with our ignorance of the steps of the migration of plants and the absence of chronological data of the distribution of plants, appear to me to render this view unsatisfactory, and like the preceding to involve supererogatory creative acts, not according with the simplicity so characteristic of the laws under which the natural forces are made to act.

Admitting then this hypothesis of creation at specific centres, let us see what are the consequences to be deduced from it. We may suppose the species to have been pretty equally apportioned to equal latitudes over the globe, and located in the different regions according to their physiological characters. As they spread they would become mingled together like the circles from raindrops in a pond, and complex conditions at any particular point after the lapse of a long space of time would be explicable by the influence of external agencies upon the individual species, and their influence upon one another, as exerted by those of more vigorous growth bearing down and repressing the growth and dissemination of the more tender.

How far the facts we meet with agree with such a view is the test by which the hypothesis must be tried, and the researches of botanical geographers appear to give a favourable verdict in the present state of the inquiry. It is necessary, however, to be very comprehensive in our examination of the phænomena, since the interfering causes are so numerous that false conclusions may readily be drawn from the investigation of small areas, or from too little weight being attributed to what often seem to be very trivial external agencies.

It would naturally be supposed that the dissemination of species would go on most freely and copiously over large continents, and that the present condition of nature would exhibit more instances of peculiarity in the floras of given regions in proportion as those regions were cut off by some natural barrier, such as mountain ranges, oceans, &c., from other lands; moreover, that islands would present a more characteristic vegetation than tracts of equal extent on a continent, and would approximate in their characters most closely to the main lands nearest to which they were situated. Such is actually the case.

The continents lying in the northern hemisphere are almost continuous in their most northern regions, and we find the greatest agreement in their vegetation at those points. In proportion as we pass south, the Old and New Worlds become more and more unlike, while the great continent of Australia lying off, in such different parallels, from the tropical and sub-tropical parts of Asia, is eminently singular and peculiar in the character of the plants which inhabit it. The opposite sides of large continents, even where there is continuity of land between the regions, display great diversity of vegetation under almost similar climates, while mountain ranges, interposing climatal barriers, cut off closely adjacent countries. Islands lying in the Atlantic exhibit peculiar plants intermixed with others which have come from Africa, Europe, or America, and in proportions agreeing with the contiguity and the identity of climate, modified perhaps by peculiarly constituted diffusing agencies of winds, ocean currents, or birds, in particular cases.

The most important question then, for the explanation of the existing states of distribution, under the hypothesis above proposed, is the investigation of the characters of those natural agents, which, while the conditions considered in the foregoing chapter *suffer*, are actively engaged in *effecting* the dissemination of plants.

The fruits and seeds of plants are so organized as to be less perishable than any other parts of the structure, and experience has shown that they possess in general a very striking power of resisting external injurious agencies. The cases are comparatively rare where it is necessary that a seed should be committed to the earth the instant it falls from the parent tree, while on the other hand a vast number both of seed-vessels and naked seeds are provided with peculiar apparatus for the purpose of ensuring their removal from the vicinity of the plant by which they are produced. The existence of wing-like processes, of hairy or feathery crowns, or the minute size, rendering their weight almost inappreciable, places them at the disposal of every current of air ; and when we consider the force of the wind, when we recollect the size and weight of objects that are taken up and carried away even by mere gales, the fact of the occurrence of stormy, blowing weather, to a greater extent than usual, at the season when fruits and seeds are ripe and falling, would lead us to wonder rather that the generality of plants are not more widely disseminated through the regions where the climate suits them, than that they do actually spread over large areas. It is clear that such seeds as those of the poplar, the fruits of most of the Compositous plants, such as the dandelion and the groundsel, with their feathery wings, might readily be carried across Europe by a powerful autumn gale blowing steadily in one direc-

E

tion, and that therefore it is only what we ought to expect when we find such plants distributed sporadically, that is, universally, without any evident centre being discoverable in this late epoch of their gradual dissemination. We have a striking instance of the facility with which such plants spread, in the *Erigeron canadensis*, a species of flea-bane, a plant imported into Europe since the discovery of America, and now a common weed on the continent of Europe.

When we take into account exceptional winds, occurring in storms and hurricanes, we see at once that there can scarcely be any seeds capable of resisting their transporting power, that almost every kind may be conveyed along continuous tracts of land, and that very many kinds may be caught up and carried along with sufficient force to account for their presence in islands, or on continents separated by considerable arms of the sea : the existence of intermediate chains of islands between distant continents would account for a certain agreement in the floras of the two most distant points, an interchange taking place by a more gradual operation, a succession of short transportations from one link of the chain to another.

Currents of water, of rivers and of the sea, are in like manner very efficient instruments for the dissemination of plants, in particular of many kinds which are less at the mercy of the winds, kinds having firm and solid coverings to their seeds or fruits, which, as adding to their weight, render them less liable to be carried away by the winds, but on the other hand enable them to resist the decomposing action of even salt water for a considerable time.

Plants growing by the water, near the sources of rivers, will naturally acquire in time an equable distribution

down the course of the stream so far as the climatal conditions will allow. Alpine plants are in this way frequently brought down even into an unlike climate, and
a curious phænomenon is frequently noticed here, when
the species is one that cannot accommodate itself thoroughly to the lowland climate. It germinates and
flowers, but perhaps seldom ripens seed, so that the continued presence of the species in the lowland station is
not the result of a firmly settled colonization, but is sustained by the advent of a succession of emigrants from
the original source, continually brought down to dwindle
and die out upon this outpost. But when the species is
of a more versatile habit, the colony becomes established,
and forms a new centre whence the distribution takes
place over the surrounding land.

Winds and aqueous currents co-operate on a large scale
for the transport of fruits and seeds. The wind acts like
the Canadian *lumberer* or the Tyrolese wood-cutter, who
fells the timber of the forests growing near the stream,
brings it down to the banks and commits it to the current to be carried downward : the wind, stripping the
ripened autumnal vegetation, rifles it of its fruits and
seeds, and then casts a large portion of its gatherings
into the stream, to float downward until, caught by some
projecting bank or diverted by some eddy, it effects a
lodgement, or, steering clear of all such obstacles, drives
out to sea, not unfrequently to find a new home on some
comparatively distant island. Continued gales of wind
blowing off the land must act in the same way to cast
large quantities of seeds into the ocean, and the tides and
currents convey a notable proportion of them to almost
incredible distances. Mr. Brown found that 600 plants
collected about the river Zaire in Africa included 13

species natives also of Guiana and Brazil; these species mostly occurred near the mouth of the river Zaire, and were of such kind as produced fruits capable of resisting external agencies for a long time. Dr. Hooker states, as the result of the examination of a large number of insular floras, that the family which in these contains the greatest number of species common to other countries is that of the Leguminosæ, where the firm seed-case, the pod, forms a very efficient protection to the seed, and is from its shape and buoyancy exceedingly well adapted for water conveyance.

The animal creation also contribute in a variety of ways to the dissemination of plants. Many fruits are provided with coverings armed with minute hooks or other prehensile contrivances, by which they catch any passing object, and then, torn off from the parent plant, are conveyed to distant spots. The seeds which form a large proportion of the food of so many animals frequently pass through the intestines uninjured; the dung of the horse always contains abundance of seeds of hay grasses and oats, which, having escaped comminution by the teeth, are voided still capable of germination; and hence, as observed by Linnæus, the frequency of the occurrence of weeds in well-tilled lands, sown with clean seed, especially when it has been manured with fresh dung. Birds are still more important agents in this way, from their more locomotive habits, in particular the migratory kinds. Many of the seeds swallowed by them pass away undigested, protected by their horny coats; and among the fruit-eating birds, the pulp, the more nutritious part, being assimilated in their stomachs, the indigestible seeds pass uninjured and in a state which causes them to germinate more quickly than usual. The seeds of the pulpy fruits are frequently provided with horny coats, and in the natural condition the softening of this and the stimulation of the germ within are dependent in some degree on the decomposition or rotting of the pulpy envelope. Should such a fruit fall in a dry place, its seeds may remain long dormant. The imperfect process of digestion which operates upon the coats of such seeds when the fruits are devoured by birds, performs exactly this office, and moreover being deposited among the stimulating excrementitious matter, they germinate quickly. It is the common practice in some parts of England to raise seedlings of the hawthorn by feeding geese and turkeys on the haws and sowing their dung ; by which means the plants are obtained the year after the seed is ripe, while in the natural state they seldom germinate till the second year. There is no doubt that the distribution of pulpy fruits in general, which are usually heavy and not liable to be conveyed by winds or aqueous currents, has been and is still greatly dependent upon birds. Sir Charles Lyell in his ' Principles of Geology' gives a graphic picture of the combined effects of the influences of which we have been speaking :—

" The sudden deaths to which great numbers of frugivorous birds are annually exposed must not be omitted as auxiliary to the transportation of seeds to new habitations. When the sea retires from the shore, and leaves fruits and seeds on the beach, or in the mud of estuaries, it might, by the returning tide, wash them away again, or destroy them by long immersion ; but when they are gathered by land-birds which frequent the sea-side, or by waders and water-fowl, they are often borne inland ; and if the bird, to whose crop they have been consigned, is killed, they may be left to grow up far from the sea.

Let such an accident happen but once in a century, or a thousand years, it will be sufficient to spread many of the plants from one continent to another; for in estimating the activity of these causes, we must not consider whether they act slowly in relation to the period of our observation, but in reference to the duration of species in general.

" Let us trace the operation of this cause in connexion with others. A tempestuous wind bears the seeds of a plant many miles through the air, and then delivers them to the ocean ; the oceanic current drifts them to a distant continent; by the fall of the tide they become the food of numerous birds, and one of these is seized by a hawk or eagle, which, soaring across hill and dale to a place of retreat, leaves, after devouring its prey, the unpalatable seeds to spring up and flourish in a new soil.

" The machinery before adverted to is so capable of disseminating seeds over almost unbounded spaces, that were we more intimately acquainted with the economy of nature, we might probably explain all the instances which occur of the aberration of plants to great distances from their native countries. The real difficulty which must present itself to every one who contemplates the present geographical distribution of species, is the small number of exceptions to the rule of the non-intermixture of different groups of plants. Why have they not, supposing them to have been ever so distinct originally, become more blended and confounded together in the lapse of ages ?"

The agency of man in the dissemination of species becomes an exceedingly important element of the question in investigating the botanical geography of long-inhabited lands. In the case of cultivated plants introduced into countries within historical periods, there is no difficulty ;

it is a known fact that the potato came from America, that the coffee-plant was carried from the Old World to the New, and so of many others. But even among cultivated plants we meet with difficulties which are at present inexplicable. The native countries of wheat, and most of the other cereal grains, are unknown, as is the origin of the banana of the tropics ; and when we recollect that few products of cultivation can be collected without intermixture of the seeds of wild plants, that is, of weeds, and when we know that modern armies have carried corn and cultivated vegetables from one extremity of Europe to another, we see a large opening for the introduction of Eastern plants into Europe by the great expeditions of the ancients, the conquests of Alexander, of the Romans, and afterwards by the Crusaders. We know that many plants have become naturalized in England, after introduction with seeds of corn and fodder plants, even within recorded periods ; and the addition of new forms, by ballast cast upon the sea-shore or banks of maritime rivers, is an operation which is going on now more rapidly than ever. The *Œnothera biennis*, the *Mimulus luteus* and *Impatiens fulva*, are examples of undoubted North American plants which have now taken a firm stand in various parts of Britain.

There are, moreover, a certain number of plants which seem to accompany man wherever he goes, and to flourish best in his vicinity ; thus the docks, the goose-foots, the nettle, the chickweed, mallows, and many other common weeds, seem to be universal, though unwelcome companions to man, dogging his footsteps, affording by their presence, even in now deserted districts, an almost certain index of the former residence of human beings on the spot.

For the diffusion of plants, then, we have agents enough and to spare, and the point hardest to decide is generally one arising from peculiar limitations of the areas of plants. We find one plant existing over a large area, others only upon small ones; some with well-marked foci or centres of distribution, others equably diffused; others again occurring in two or more isolated localities in which they have apparently independent areas of distribution; finally, we find large tracts of country covered by masses of vegetation, consisting of a small number of species, while other small regions yield a remarkable variety.

Here we have a series of problems of very different nature from the foregoing, and which require a very comprehensive investigation for their solution. It is this part of the subject which at present forms the chief field of inquiry in botanical geography.

The causes of limitation of species may be divided into several classes:—1. Those connected with the original creation; 2. those arising out of the special character of plants and their influence upon one another; 3. those dependent on varying climatal influences during the period of the plant's existence; 4. the influence of geological changes during the same period; 5. those produced by animals; and, 6. those produced by the agency of man. The causes of the varying climatal influences will be most properly distributed under the fourth and sixth heads, since they must either have arisen from changes of elevation of the earth's surface, or from modifications of the condition of moisture or exposure through the operations of man, by such means as the felling of forests or draining of large tracts.

With regard to the phænomena of the first class, it is matter of inquiry whether all existing species were created contemporaneously, and whether in greater numbers on particular spots, or equally over the globe. To the first question it is impossible to give any decided answer; the only positive evidence we could have would be the appearance of some altogether new species in a spot which was thoroughly well-known and investigated. No such case is known, and, moreover, in the present condition of knowledge, it would be exceedingly difficult to authenticate one. Still it is possible to imagine that species have been successively created during geological epochs, as well as at the commencement of them, of which we have a convincing demonstration from the investigation of fossil plants. And some botanists have suggested that the peculiar and otherwise unaccountable limitations of certain species depend upon the later creation of some ; and that these, finding certain tracts too fully occupied by other already widely-diffused species, have been thus prevented from acquiring that degree of diffusion which their other characters would lead us to expect. This, however, is purely a point of speculation in our present condition of knowledge, and probably will never be absolutely decided.

With regard to the second question, Have plants been created in crowds at certain points, or equally?—we again are unable to answer positively from experience, and our conclusions must be drawn from calculation of the value of the modifying agencies to which the vegetation of the various regions has been subjected for indefinite periods.

Islands generally appear richer in *peculiar* species than tracts of equal size on continents ; but it is evident that this must be so, and the only correct way of analysing the question would be, if possible, to find an entire

flora, well-defined and peculiar, of a large continent, as
of Australia, and then to compare the number of peculiar
plants given by a small island with the number obtained by dividing the whole number of plants in the
flora of the continent by the figure indicating the number
of times it exceeded the small island in size. Such a
process cannot be adopted with hope even of approxi-
mately true results while such vast portions of the surface of the large continents remain unexplored, and
Europe is manifestly unfit to serve as the basis of the
calculation, its flora being so demonstrably compound in
its origin. It seems simplest and fairest to assume that
species were at first equably diffused in equal latitudes,
or perhaps in isothermal parallels; it is known that
species increase towards the equator and diminish towards the poles, a natural arrangement in accordance
with the conditions of climate, and one most probably
original and primary, since it is unreasonable to imagine
that plants have been created merely to be speedily destroyed by inevitable, pre-ordained influences.

The second class of causes are highly important, and
give some of the most remarkable peculiarities to the
distribution of plants. A striking example is afforded
by pine forests, where the surface of the ground is almost
bare of vegetation, and brown with the fallen leaves and
branches of the trees, adding a still more gloomy character to that derived from the tall and regular trunk,
and dark, opake foliage. Large tracts of country in
Britain and the north of Europe are overrun with furze
or with the heath, which admit of few companions among
them; and all gardeners are familiar with the power
which some plants have of destroying others.

Those with large and widely-extended roots take away

the nourishment of their more delicate neighbours ; tall plants overcrowd and intercept the light from the humbler kinds ; trees destroy herbs with the drip of rain from their foliage.

Thus, supposing a number of plants, varying in habit, were created in a given tract, those most vigorous would necessarily acquire a more extended area than the others ; and the existence of a full vegetation of vigorous habit may form a barrier to the gradually extending area of some neighbouring creation, stopping the extension at that point almost as completely as a climatal barrier.

Thus, when vegetation in general has attained a certain degree of diffusion, the circumstances arising from climate and special character may form a complete barrier to its further extension, and the continuance of the latter, the increasing power gained by the stronger plants over the weaker, may, it is clear, lead to the extinction of species at particular places. Therefore it by no means follows, when we find a species distributed in a series of isolated patches, that those detached areas are the result of so many distinct creations, since they may be all descendants from one stock, the progeny of which have spread themselves along a line in a certain direction, and have subsequently been cut off by the intersection of more vigorous species, whose areas have at intervals spread across, interrupted and separated it into patches by their greater strength and luxuriance. The principal causes which prevent the total destruction of the weaker by the stronger are the peculiarities of habit ; the preference by particular species of some peculiar kind of *station*, as it is called. Some plants flourish in meadows, some in swamps, others in the clefts of rocks, others on the sea-sands, &c. Consequently, where the plants find their most suitable conditions, they take absolute possession, but leave the other kinds of station quite free to other species. Plants may also protect and foster one another; some species will only grow in the damp shades of woods, others require the shelter and protection of hardier kinds against the northern blast; and the spiny arms of one species may prove the guardians not only of its own tender buds, against the attacks of animals, but of a number of tender little plants which nestle under its shelter. On our commons, for instance, the blue-bell, the milkwort, the dog-violet, and a number of other delicate flowers are found cowering under the sides of the furze-bushes, while the surface of the soil around is eaten close by the cattle.

So long as it was imagined that the surface of the globe had been at rest ever since the creation of the organized beings now inhabiting its surface, no one would seek to explain anomalous cases of distribution by geological evidence. But now that we know great changes to have taken place since the first appearance of many animals at present existing, and that we are uncertain as to the precise epoch of the creation of existing vegetation, it not only becomes admissible, but even requisite, to inquire how far changes in the arrangement of land and water, and of the elevation of the former, during the period of the existence of the vegetation of the modern period, may have affected the distribution of species, and contributed to produce those conditions which we now meet with.

Even the quiet and regular progress of nature is a continuous operation of changes. Water rises up in vapour from the sea; cooled by the air-currents or by mountain-summits, the aqueous vapour falls as rain; the streams, which drain the watersheds or declivities of mountain-ranges, gathering strength as they proceed, at last become great rivers, carrying along in their course the soil they eat out from the surface they pass over; sometimes they deposit much of this in lakes occurring in their course, often on their own shores in the tortuous windings of the lower parts of their course, or they cast it forth into the sea to form banks, or even finally islands at their mouths. The sea, too, undermines and washes down its boundaries in one place, and casts a tribute of new soil upon the shores of another.

In the southern hemisphere new islands, the combined work of coral polypes and submarine volcanic action, rise from time to time from the bosom of the waters; while the active volcanoes pour torrents of destructive lava over tracts of vegetation, and the blown sands of the deserts, or of certain sea-coasts, gradually invade and cover up the more fertile soil adjoining them.

Such constant change, such an unceasing destruction and reparation of surface brings with it the destruction of masses of vegetation and the production of new areas to be repeopled, often with a vegetation of totally different habit, fitted to the altered physical conditions which have supervened. It is evident that such phænomena must be highly important in modifying the general distribution of plants.

But if we suppose, as we have a right to do, from the ascertained facts of geology, that greater changes, involving much more striking alterations in the condition of the earth's surface, have taken place since the present flora of the globe was created, we are led to contemplate the possibility of a far greater amount of influence of this kind having contributed to produce the existing state of things.

Let us see how some of these geological phænomena would affect the distribution of plants. The operations would be of two kinds : first, to arrest or to favour the dissemination of species from their centres, and subsequently to isolate portions of a continuous area by partial destruction, or to destroy entire areas ; and secondly, to change the physical conditions of regions, altering the climate, or even overspreading them with deposits of soil of different constitution.

The interruption of the continuity of the earth's surface, subsidence of land, producing an archipelago of islands in place of a continent or cutting off two continents by a wide strait, would isolate portions of an existing area of distribution, as would in like manner the elevation of a mountain-ridge across it, and from such causes the mode of distribution and the locality of the original centre would be in a great degree hidden and rendered impossible of explanation without geological evidence. The gradual elevation of points of land above the sea, contemporaneous with the formation of new tracts spreading out from their bases, if accompanied by changes of climate resulting from a widely extended alteration of the general distribution of land and water, producing a higher degree of temperature in the latitude, would isolate species upon the elevated points, where the elevation would continue to them their originally cooler climate, while the new tracts would be invaded by the inhabitants of other regions originally warmer. Thus a group of islands in an icy sea might become the mountain-summits, clothed with Alpine plants, scattered over a temperate continent.

The possible changes of climate as well as of geographical configuration from comparatively slight causes, is much greater than would at first be imagined, and this has been ably illustrated by Sir C. Lyell.   Thus he says, "if the narrow isthmus of Panama were to be gradually broken down by subsidence, a few centuries would suffice to bring about most remarkable changes in the condition of living creatures in America.   Thousands of aquatic species would pass from the Caribbean Sea into the Pacific, and thousands of those before peculiar to the Pacific would find their way into the Caribbean Sea, the Gulf of Mexico, and the Atlantic.   A considerable modification would probably take place in the direction or volume of the Gulf Stream which might alter the temperature of the sea and the adjacent lands in the direction of its course.   A change of climate might thus be produced extending from Florida to Spitzbergen, and in many countries of North America, Europe and Greenland.   Not merely the heat, but the quantity of rain would be altered in many of those places, so that some species would altogether lose their natural conditions and become extinct, others would become less generally diffused, and some would be greatly benefited, so as to flourish and extend more widely than before.   The seeds and fruits of many plants would be carried by the currents in other directions, to populate localities from which they had been previously cut off, not by climate, but by the barriers of land and ocean."

In another example he shows that even a few earthquakes, not more considerable than have been observed in the limited experience of the last 150 years, producing ravines in the low sandy tracts, at present liable to suffer from them, intervening between the Sea of Azof and the Caspian, would produce an enormous change in the physical geography of Asia.   The Caspian being 84 feet lower than the Sea of Azof, the waters would pour in, fed successively by the Black Sea, the Mediterranean and the Atlantic, until not only the low sandy steppes around the Caspian would be submerged, but the waters would not stop until arrested by the high land which connects the Altai with the Himalaya Mountains. A few years, perhaps a few months, would suffice to effect this revolution.

Again, suppose a few elevations to take place in the shallowest part of the Straits of Gibraltar, where the deepest soundings are only 220 fathoms; the channel would be narrowed, and the volume of water pouring into the Mediterranean from the Atlantic would be lessened; and, as the evaporation from the surface would be the same, the level must sink, and tracts of land would be left dry around its borders. The current flowing rapidly out of the Black Sea would be accelerated to supply the want, but this would lower its level; and, supposing a series of elevatory movements completely to close the Straits of Gibraltar, the Mediterranean and the Black Sea would probably be surrounded by large, level sandy steppes like those now existing around the Caspian and the Sea of Aral. It is clear that such changes must greatly affect the distribution both of plants and animals.

If a barrier were raised across the Bahama Channel by a chain of submarine valcanoes, the Gulf Stream would be arrested. This would produce an almost incalculable change in the climate and distribution of organized beings in the northern hemisphere.

It is unnecessary to pursue this subject further here; these illustrations will suffice to show how important an element geological evidence is in botanical geography.

The influence of animals in the limitation of species is

exercised on a smaller scale, and more in detail, but yet is by no means inconsiderable. The ravages of the insect tribe are phænomena too familiar to every one to require to be dwelt upon here. We notice them in general only as affecting the success of our cultivations, but the same habits which render them such a terror to agriculturists and horticulturists, and cause them now and then to endanger the existence of large bodies of human beings, are evidently, in free nature, so many checks and modifying influences regulating the development of the various forms of vegetation. From the remarkable provision of the attachment of particular insects to peculiar plants, their operations become, in most cases, not an universal but a special limiting influence, and the strangely sudden abundance of certain insects at intervals, coming from time to time to attack their appropriate plants and acting as a periodical check, is a striking provision for the maintenance of the equilibrium among vegetable species. If the more luxuriant kinds remained unchecked, they would in many instances extirpate the weaker forms, while the continued operation of a particular insect, on a large scale, upon a particular plant would probably soon cause the extirpation of this species. As it is, time is left for it to recover its footing to a certain extent before another incursion comes to restrain it within proper bounds.

The devastations of locusts, feeding indiscriminately upon all kinds of plants, form an influence rather affecting the general vegetation of the peculiar localities which are invaded by them, than specific modifying causes of local floras. But in partial or slight inflictions of this scourge, the insects are said to select in preference particular plants, devouring all of them before they attack the remainder, and this may in some measure favour the predominance of particular species in countries where the invasion is periodical.

The influence of man in modifying the distribution of plants, exercised as a limiting influence, arises chiefly from agricultural operations of various kinds. Among those acting in a more general manner may be counted the felling of large forests, and the drainage of large tracts either of freshwater or marine marshes. There is no doubt that the humidity of the soil must be much diminished, and the general climate thus rendered drier, by the removal of large forests; yet it is asserted that the felling of forests in Europe and in North America has rendered the climate warmer in winter and cooler in summer than it was formerly, the evidence being founded in the case of Europe, on the fact that wine was made formerly at greater elevations and in higher latitudes, earlier than it is now. The climate of Tuscany is said to have been rendered less cold by the removal of wood.

Such statements as these would go to prove that the removal of wood renders the climate less excessive; but since it must render it drier, it is probable that these assertions are founded on too narrow a basis, on areas modified by peculiar local conditions, for all experience shows that dry climates are the most *excessive*, having hotter summers and colder winters, and this is probably the case with Northern Europe ; indeed the Roman writers state that not even the cherry, much more the vine, would ripen in Germany in their time.

However this may be, it is certain that the appropriate *stations* of many plants would be destroyed with the removal of forests, and new conditions of soil created for the habitations of immigrants from other regions.

But the modification of the surface so as to alter the physical condition of the soil, is by far the most important change brought about in reclaiming land to cultivation. Marshes and even lakes are laid dry by drainage as a country becomes more densely populated, and the peculiar marsh or bog plants by degrees extirpated. This process is going on very rapidly in our own country at this time; many bog plants are now lost from localities which they formerly inhabited. The banking-out of the sea, again, changes by degrees the character of the vegetation of its shores; bare sand-dunes, where scarcely a plant could maintain a precarious footing, are by degrees covered with vegetation; sandy inland wastes are rescued from the heath and furze, and made to contribute at first by Coniferous woods, such as the larch, and when the soil has become by degrees enriched, by the plants requiring a better nourishment, to the general stock of wealth; and in these changes many species are destroyed, while others naturally making their way into a fitting station, or brought undesignedly by the hand of man, grow up and wholly displace the original inhabitants.

Neither must we forget the vast importance of the careful tillage of the soil of cultivated lands by which the aboriginal inhabitants are reduced or eradicated; and if many kinds are introduced, as weeds, yet a most powerful barrier is opposed to the spread, on a large scale, of any useless plant. Thus, while it may formerly have been comparatively easy for any plant to extend from Asia across Europe in the line of its appropriate climate, none but an inconspicuous kind, or such as are extremely indifferent as to *station*, could do so now, since any remarkable plant, such as a tree or shrub, could scarcely escape destruction if unprofitable, while the diminished variety of stations in drained and cultivated countries diminish proportionately the chances of any given plant, except the cornfield weeds or way-side plants, finding the fitting conditions for its maintenance and propagation.

With these general views, involving such multifarious considerations, we here conclude this imperfect exposition of the principles which guide us in the study of Botanical Geography. The succeeding chapters will be devoted to a description of the vegetation of Europe as it exists at present, with incidental observations directing attention to changes it has probably undergone, to the bearing of the laws already enounced upon the particular cases, or to bringing into especial notice striking instances illustrative of the general facts mentioned in the foregoing pages.

The detailed reference of the single phænomena to their causes would occupy volumes instead of a few chapters, and no profession is made here to a perfection of such kind. The reader who has attentively perused these introductory chapters will find many applications of the principles, which it is impossible to enter into minutely where the mass of material that has to be dealt with is so large, and when the conclusions to be drawn must be selected from cases in which a complexity or abstruseness, out of place in a popular treatise, can be avoided. Even as it is, the frequent necessity of using the only definite names, the Latin ones, in so many instances, gives a technical aspect to the pages, yet they are introduced as sparingly as possible. The conduct of our book is in some sort like steering between the rock and the whirlpool; we have to avoid running upon the Scylla of superficiality or sinking into the Charybdis of scientific specialities; and, not altogether confident of steadiness of hand under all circumstances, we must claim that indulgence which may be fairly asked when the adventure is into a new and untried region, as is the case in this endeavour to throw the results of the scientific investigation of the distribution of vegetation of Europe into a form presentable to general readers.

CHAPTER IV.

CHARACTERISTICS OF THE COUNTRIES OF EUROPE.

IT is almost impossible to contemplate the Map of Europe, to reflect on the disparity of size between it and the other continents, and the inverse proportion of the intellectual development of its inhabitants and those of the other quarters of the globe, without being compelled to adopt the opinion that the physical conformation of the earth exercises a vast influence not only on the animal and vegetable creation, but upon the human beings who live upon its surface. It seems unquestionable that the infinite diversity of conditions in a small space, together with the freedom of communication existing between the different regions, the great extent of sea-coast, and above all the climate, compelling effort in all places, yet scarcely anywhere presenting insurmountable obstacles to the cultivation of the soil, have been very important causes in determining the position of the centre of civilization upon the globe ; and although the influence of the human mind, as its general development progresses, tends more and more to extinguish the differences, which have in all probability arisen in the earlier periods of man's history, through the preponderating effect of physical causes at that time, these causes still retain sufficient power even in Europe to mark with considerable distinctness the general characteristics of the inhabitants of countries lying under diversified physical conditions.

These same causes which in man determined the development of a single species have in like manner distinguished, as a whole, the vegetation clothing the different portions of the earth's surface in Europe, and have given, considering the comparatively small superficial extent of the continent, a very great variety of forms.   While the northern tracts run into the icy regions of the arctic circle, the high degree of temperature produced by the contiguous mass of land of Northern Africa brings straggling members of tropical families into its southern valleys; and the numerous elevated points and chains of mountains, the expanses of high table land, the wide flat plains, and the valleys of the great rivers, afford almost every variety of condition of climate, below certain fixed degrees of temperature for each region, within the limits of nearly every one of its great political subdivisions.

Europe lies between 35° and 71° of N. Lat., Lapland and the northern point of European Russia being within the arctic circle ; all the rest is included in the temperate zone, and dividing this into a cooler and warmer division at 45° N. L., we find the greater portion of the continent to belong to the former.

The general longitudinal extent may be taken from 10° W. of Greenwich to about 60° E. L. ; but a tongue of Northern Russia runs out nearly to 70° E. L. ; while if we include Iceland, the western boundary will be carried out to 24° W. L.   The portion lying between 10° W. L.

and 30° E. L. is that which presents the greatest variety of conditions, and is most remarkable for the importance both of its geographical phænomena and its political history. The vast plains of Russia lying to the east of this constitute considerably more than a third of the whole superficial extent of Europe, and unite it very completely to Asia, not only geographically, but in its natural history peculiarities, and the habits and characters of its inhabitants.

This great plain, extending from the Arctic Ocean to the Black Sea and the Caucasian chain, divided by the Ural chain and river and the Caspian Sea from Asiatic Russia, becomes narrowed gradually towards the west, on the north by the Scandinavian mountains, and on the south by the Balkan, the Carpathians, the Riesengebirge, the Erzgebirge and the Hartz mountains, enclosing within this somewhat triangular tract the Baltic Sea. To the west of the Hartz the plain widens out again between the mountains of Thuringia, the Vosges, the Jura, the mountains of Auvergne, and the Atlantic Ocean; then if we regard the mountains of Great Britain as the north-western boundary, we have here a basin, enclosing the North Sea. In this way the greater part of Northern Europe presents itself as a single great plain divided by the Danish peninsula into two very unequal portions. This great plain penetrates deeply into the higher lands in two places; in the Hungarian plains and the extensive Valley of the Danube connected with them, and in the Valley of the Rhine.

The mountains of Central Europe are distinct from the Alps, but may be regarded as forming with them a great connected tract of elevated country, the southern bases of the former being taken as lower terraces of the Alpine chain. The Apennines and the mountains of Croatia and Dalmatia constitute as it were branches of the Swiss Alps, and the Adriatic with the valley of the Po form a great level tract between these ; while the Dinaric Alps passing continuously into the Balkan and southward into the mountains of Greece, form a bond which brings all these great chains into connection, so that we may consider them as forming one large region of elevation. The Pyrenees are completely separated from this, but stand in immediate connexion with the mountains of the Spanish peninsula.

Europe may therefore be divided into four great parts :—

1. *A large S.E. Highland,* the Swiss Alps, the mountains of Central Europe, the Apennines, the Dinaric Alps, the Balkan, and the mountains of Greece.

2. *A smaller N.W. Highland,* the Scandinavian chains, with which those of the British Islands may be naturally grouped.

3. *A smaller S.W. Highland,* the Pyrenees with the Spanish peninsula.

4. *A great plain,* bounded by these three Highlands, the Ural Mountains, and the Atlantic Ocean.

Besides these, there exist certain smaller elevated regions, namely the mountains of the Crimea, which should probably be considered as belonging to the Caucasus, the mountains of Iceland, and a few smaller groups of islands.

The mountain masses of Europe present great variety not only in the height, but in their horizontal extent and direction. The following general arrangement will facilitate the comprehension of the part they respectively play in the different regions.

Grouped according to mass, we have:

1. Mountains of the first order.—The Scandinavian Alps, the Swiss Alps, the Apennines, and the Carpathians.

2. Mountains of the second order.—The Balkan, the Dinaric Alps, the Greek mountains (considered as a whole), the Icelandic (in like manner), the Pyrenees, the Asturio-Gallician mountains, Guadarama, Sierra Nevada, the Cevennes, the Jura, the Scotch mountains (collectively), Sierra Guadeloupe, Sierra Morena, the Vosges, and the Sicilian mountains.

3. Mountains of the third order.—The English mountains, Sardinian, Auvergnese, the Böhmerwald, the Corsican mountains, the Black Forest, the Suabian Jura, the mountains of the Crimea, Sierra Monchiqua, the Riesengebirge, the mountains of Ireland and of the Feroës.

Grouping them according to the elevation of the highest peaks, we get the following classes.

1st Order.—Mountains which attain a height of 15,000 feet: the Swiss Alps.

2nd Order.—Mountains with the highest points about 11,000 feet: Sierra Nevada, the Pyrenees, and Etna.

3rd Order.—Mountains between 8000 and 10,000 feet: the Apennines, the Corsican mountains, the Carpathians, and the Balkan.

4th Order.—Mountains between 6000 and 8000 feet: the Guadarama, the Scandinavian Alps, the mountains of Greece, the Dinaric Alps, Sicilian chains (excluding Etna), the Icelandic mountains, and the Cevennes.

5th Order.—Mountains between 4000 and 6000 feet: the mountains of Auvergne and Sardinia, the Jura, the Riesengebirge, the Majorcan mountains, the Crimean chains, the Black Forest, Minorcan mountains, the Vosges, and the mountains of Scotland.

6th Order.—Mountains under 6000 feet: the rest of the European mountains of which the height is known.

The average height of the chains would in some instances form a different set of groups, but these have not been so clearly ascertained and therefore cannot be used so safely as a guide.

The directions of the great chains of elongated form are :—

East and West: in the Swiss Alps, the Balkan, the Pyrenees and the Spanish mountains collectively.

North and South: in the Scandinavian Alps, the Cevennes, the Vosges, the Black Forest, and the mountains of Corsica and Sardinia.

North-west and South-east: in the Apennines, the Dinaric Alps, the Riesengebirge, and the chains of the Böhmerwald.

South-west and North-east: in the Jura, the Suabian Jura, and the mountains of Scotland.

The Auvergne mountains and the Hartz have no remarkable difference of length and breadth.

Groups of mountains occur in Europe in Greece, Iceland, Sicily, Ireland, and in the mountains which are situated on the curved projection of the Carpathians.

Etna, Hecla, Montserrat, Vesuvius, and Gargano are isolated mountains of considerable height situated in plains.

The great plains of Europe are—1. the east European; 2. the north European; 3. the Hungarian.

Of the extensive elevated plateaux, the Spanish is the largest and highest; the Bavarian next.

The peculiarities of the temperature of Europe, and

especially the great difference between the eastern and western portions, arising from the maritime position of the latter, and its subjection to the influence of the more general causes diverting the isothermal lines so far to the north in the Atlantic, have been discussed in a former chapter. It only remains now to furnish a more definite account of the actual conditions of the temperature within this continent; and this may be made most clear by reference to the accompanying Map, illustrating the mean temperature of July and January, in Europe, by means of isothermal lines drawn so as to connect all places together in which the average heat received in those months is equal.

The blue and green lines denote the winter or January temperatures, given in degrees of Fahrenheit. The bright blue lines mark isotherms with a temperature below the freezing-point, as is seen by the marginal figures on the left-hand side of the Map. A dotted line coloured with indigo passes through all those places where the mean temperature of January corresponds to the freezing-point. To this succeed the green lines which indicate temperatures rising successively above the freezing-point of water.

The shaded purple lines connect all the places which have what is called the normal temperatures of January. The figures on the left side of the 20th meridian of longitude are connected with these. These figures signify the calculated mean temperature of the places all round the globe situated on the parallel at which the figures stand. Now the blue and green lines *rise* from the thermic normal of January in America and continue above it until they descend low down in Southern Russia; within the line in America and in S. Russia the January isotherms fall; from this it follows that the temperature of January is

The original colors of this map are represented by the following numbers: 1-bright blue lines, 2-a dotted line with indigo, 3-green lines, 4-shaded purple lines, 5-yellow lines, 6-red lines.

above the average height for the latitude all over the N. Atlantic and over the greater part of Europe; while in the continent of America and in S. Russia the winter is colder than the average.

The yellow lines indicate the lines of equal temperature in July, the summer heat diminishing, in the degrees indicated by the respective figures on the right-hand margin, from south to north, with tolerable uniformity in the south of Europe, but with increasing irregularity northward; and, what is especially to be noticed, their course is contrary to that of the winter lines; so that places having intense cold in January are found to have great heat in July; the greatest equality of temperature, the least variation in the seasons, being seen to exist on the western coast, subject to the influence of the Atlantic Ocean.

The red lines correspond to the purple ones. They are the thermic normals for July, and cut all the points which have the average temperature of the latitude (as observed all round the world) in that month. The normal temperatures are marked on the right side of the 20th meridian. All places to the east of that in Europe are hotter, and all to the west cooler than they should be, in July. This last condition scarcely occurs in Europe, only in Ireland and Scotland, at the extreme point of N. Spain, and in the Feroë islands and Iceland. The summer heat of the various parts of Europe is seen by examining the yellow lines and finding between which any tract lies.

With regard to the quantity of rain that falls in Europe, a few general statements will suffice; more minute information may be obtained by consulting the Physical Atlas of Berghaus.

The greater part of central and all the north-east of Europe lie in the province of the autumnal rains; here the annual fall is comparatively light, varying from 15 to 25 inches in the west, greatest of course in mountainous districts, and diminishing eastward, being below 15 inches in eastern Prussia, Poland, and the great plains of Russia.

Norway, the British Isles, the west and south of France, the Spanish peninsula, Italy, Hungary, and the north of Greece, lie in the province of prevailing autumnal rains. The greatest fall is here found in the west, and on the western faces of the mountains. The average amounts to from 25 to 30 inches over the greater part of the province; but in the S.W. of Norway, the N.W. of Scotland, the S. of Ireland, the S.W. of England, the N.W. of France, and in the W. of Portugal, it rises above an average of 35 inches, as also in the western slopes of the Pyrenees, and the whole length of the Swiss Alps. A remarkable exception is met with in Spain on the high table-land of Castile, having a mean elevation of 2300 feet, over which the average fall of rain is below 10 inches.

The S.W. extremities of Portugal and Spain, Sicily, the south extremity of Italy and Greece lie in the province of prevailing winter rain. All these are comparatively dry, except some points on the coast of Spain.

Eternal snow is met with in Iceland and the Scandinavian Alps, on the Balkan, the Swiss Alps, the Pyrenees, and on the Sierra Nevada.

The Carpathians, the Apennines, Etna, and the Corsican mountains, only just border on the snow-line. In the extreme north of Europe this line lies about 2200 feet above the sea, and from the appearances of Etna, at about 10,500 feet in the south. It is lower in proportion to

the proximity to the Atlantic. The following are some
of the principal heights ascertained :—

	Feet.	
North Europe	2286	Von Buch.
Hammerfest, 70° N.L	2629	,,
Iceland, 63°·5 N.L.	2719	Hisinger.
Norway, 63° N.L.........	5167	,,
——, 62° N.L.	5271	,,
Swiss Alps.............	9063	Various authors.
——————	8744	Wahlenberg.
Apennines, 42° and 43° N.L.	9503	Schouw.
Medium of Pyrenees	8936	Humboldt.
Mont Perdu	8316	Parrot.
Pic du Midi	9625	,,

Glaciers are met with in Iceland and Scandinavia, and
on the Swiss Alps. Indications of them occur in the
Pyrenees and Carpathians.

The transparency of the atmosphere diminishes from
south to north, as also from east to west, for the inland
tracts are much less subject to mists and clouds than the
coast regions.

————————

We now proceed to examine separately the natural
provinces into which we have found Europe divisible.
They are those laid down by Prof. Schouw in his 'Sketch
of the Physical Geography of Europe,' to which work we
are also indebted for much assistance in drawing up the
descriptions of the various regions.

Sect. 1.

The Scandinavian Peninsula.

THE countries of Norway and Sweden constitute a large elongated peninsula, connected at the north by a rather broad isthmus with the mainland of the continent. Its length from north to south and from N.E. to S.W. amounts to about 240 geographical miles, the breadth varies from 50 to 95 geographical miles. Almost surrounded by water, the western coast is indented by numerous narrow, deep and irregular bays (*fiords*), while countless rocky islets lie scattered before it ; the east coast, on the contrary, washed by the Baltic, presents scarcely any bays of importance, and but few and inconsiderable islands.

The greater portion of the peninsula is occupied by a chain of mountains, the Scandinavian Alps, extending from the north to the south-western parts, from 71° to 58° N.L. These stretch first from N.N.E. to S.S.W. under the name of the Lapland Mountains and Riolen, then as Dovrefield from E.N.E. to W.S.W., and more to the south from N.N.E. to S.S.W. under various denominations, Langfield, Sognefield, and Hardangerfield, to the vicinity of the west coast. At the northern extremity the mountains slope gradually down to the White Sea ; on the south-east a natural boundary is formed by a chain of large lakes, little elevated above the level of the sea. There is no proper ridge along these mountains ; the chain is tolerably even, and presents but a slightly undulating line in its greatest elevations, which are occupied in a great extent by high mountain plateaux, subject to a rude climate and mostly uninhabited, frequently rendering the journey from one side of the mountains to the other a matter of much difficulty.

On the east side the mountains sink very gradually towards the Gulf of Bothnia. The northern portion passes almost imperceptibly into the low plains of Lappmark ; further south, side chains of considerable size cut the country into large cross valleys, such as Herjedal, Osterdal, Gulbrandsdal, Valders, and Hallingdal ; or the great mountain mass projects towards the east, and forms high plateaux or smaller groups of mountains, as in Tellemark.

The west side, on the contrary, is very abrupt ; deep narrow fiords take the place of the valleys, and are scarcely visible from the high plateaux above until the very margin is reached, whence the traveller looks down as it were over a deep precipice, at the foot of which, on the seacoast, a few little patches of cultivation denote the presence of inhabitants. The mountain-chain is here often so steep, that intercourse along the coast is chiefly carried on by water. The course of the rivers is greatly dependent on these differences of inclination ; the largest flow towards the east, while few are met with in the west.

Lakes of considerable magnitude exist both on the mountain plains and at the eastern foot ; none of any size are found on the west. Some very large waterfalls occur in the Scandinavian Mountains, among which are the Rinkan-Fossen or " Smoking-Waterfall," about 800 feet, and the Borring-Fossen, about 900 feet in perpendicular depth. Taken as a whole, the mountains are higher in the south than in the north ; the great body of the chain attaining 4500 feet in Hardanger and Sognefield,

3000 in Dovre, and about 2500 in the mountains of Lapland.   The highest known peaks are :—

	Eng. feet.
Gousta, in Tellemark	6,163
Justedalsbrae	6,375
Skagestoltind, in Sognefield	8,044
Lodalskaabe	7,243
Sneehatten, in Dovrefield	7,543
Syltop	5,851
Sulitelma, in Lapland	6,168

The Scandinavian mountains, excluding the lower ones lying in Southern Sweden, cut off by the lakes mentioned above, are almost wholly composed of primitive rocks, gneiss constituting the chief ingredient.   Limestone here plays but a secondary part.

The contrast of the climates of the east and west coasts has been alluded to in a former chapter : the former, having the continental condition, offers dry air, spring rain, warm summers, and cold winters ; while the latter has the insular or maritime climate, with its moist, cloudy atmosphere, frequent rain, mild winters, and cold summers.   Those places on the south-east side of the mountains, Christiania, Stockholm, and Upsala, give almost equal annual mean temperature, about 42° Fahr. ; while Ullensvang, in about the same latitude, on the west side, has a mean of 44°.   The winter is about 5° warmer in the east, but the summer rather more than 1° colder. Comparisons between Drontheim and Umea, of the North Cape and Enontekis in Lapland, afford similar contrasts. At the North Cape the winter is only about 2° colder than in Stockholm, while the summer heat is not greater than that of the autumn in the latter city ; in Enontekis, on the other hand, the summer is more than 11° warmer, the winter 21° colder than at the North Cape.

The following table gives a summary of the temperatures throughout the Scandinavian peninsula :—

	Latit.	Ann. Mean.	Winter. Dec.–Feb.	Summer. June–Aug.
Stockholm	59½°	42°	25°	62°
Ullensvang	60	44	30	61
Drontheim	63½	40	23	59
Umea	64	35	14	57
North Cape....	71	32	23	43½
Enontekis	68½	27½	2	54½

(1350 ft. above the sea.)

The result of sixty-eight years' observation in Stockholm gives the extreme of heat at 95°, the extreme of cold at 26½° below zero Fahr. At Umea and Enontekis mercury has been frozen, which indicates a cold of more than 35° or 36° below zero.

The seasons of Scandinavia, like the other parts of the extreme north, are very peculiar. The summer is short, but dry and warm on the east side and in Sweden, while the cold is very intense in winter, but the surface is then for a long time protected by a covering of snow. As the sun acquires power, this melts very rapidly, and the progress of vegetation is exceedingly active, the snowy tracts often changing to verdant and flowering meadows almost in a few days, since the solar influence is of such long duration, there being no absolute night in any part of Scandinavia at midsummer, the sun shining constantly in the more northern regions ; and thus a summer month brings vegetation more forward than six weeks will do in the middle latitudes of the northern hemisphere.

The annual amount of rain in Stockholm and Westeras is $17\frac{1}{2}$ inches, while in Bergen it is $77\frac{1}{2}$. Although these numbers should not, probably, be taken as indices for the entire east and west sides of the mountain chain, they lead, in conjunction with the theories based on common observation, to the conviction that the quantity of rain that falls on the west side of Scandinavia far exceeds that on the east. The abundant vapours which rise from the ocean, and the frequent precipitation which must occur when the air, loaded with watery vapour, is carried by prevailing winds against the high and steep mountains, sufficiently account for the damp, rainy, and cloudy climate of the west. The equable temperature, the little difference between summer and winter, is attributable to the influence of the ocean, always tending to equalize temperature, and not frozen in this latitude. Its influence is cut off from the east side by the mountain ranges.

Here, as in all other places, the temperature diminishes as the mountains are ascended ; and at a certain elevation the decrease is so great that the snow never melts. The " snow line " at about 2300 feet at the North Cape rises to 5300 in Southern Norway ; and is throughout lower on the west than on the east, because the snow does not melt so readily during the damp, cloudy summers, as in the clear atmosphere of the east. Glaciers extend from the great snow-fields of Folgefond and Justedalsbrae through the valleys down even into the cultivated fields : at the foot of these occur the *moraines*, heaps of stone, rubbish and earth, accumulated by gradual slips ; and the water flowing from these glaciers is rendered milky by the quantity of earthy matter suspended in it. In the northern part of the peninsula the glaciers reach the sea level.

The peninsula is abundantly wooded. The forests are

G 2

principally composed of spruce-firs (*Abies excelsa*), pines,
or Scotch firs (*Pinus sylvestris*), and birches (*Betula
alba*). The oak (*Quercus Robur*) does not reach beyond
Sondmor (63°) in the west, and only as far north as
Gefle (60½°) in the east: the beech is met with only at
Laurvig (59°) in Norway, and in Sweden not beyond the
great lakes forming the southern boundary of the mountains.

Of the three prevailing trees the birch goes almost to
the North Cape (70–71°), the Scotch fir to Alten (69–70°).
The spruce-fir goes only to Kunnen (67°) on the west
side; but on the east, where the climate suits this tree
better, 2° farther south. The hazel, on the contrary,
flourishes better on the west, where it occurs as far up as
Helgeland (65½°), while on the east it is not found beyond Angerman-Elf (63°): the lime also (*Tilia europæa*)
advances farther north on the west than on the east,
occurring at Oreland (64°) on the former, and limited by
the 63° on the latter; while the elm (*Ulmus campestris*)
attains the 63° on both sides.

Ascending the mountains, the pines and firs are lost
at a certain height, the birch alone forms the woods;
then these gradually disappear, giving place to the dwarf
birch (*Betula nana*) and a few species of willow, forming a low bushy vegetation. With these come many
little perennial plants, the alpine plants, with their large
and bright-coloured flowers, as well as numerous lichens,
among which the Iceland moss (*Cetraria islandica*) and
the reindeer moss (*Cenomyce rangiferina*) abound, and
cease with all vegetation at the snow line.

Three zones may thus be distinguished; those of the
conifers, of the birch, and of the alpine plants. In the
south of the mountains the conifers ascend to 2800 feet,

and the birch to 3500; in the north of Lapland, the former to 700, and the latter to 1500.

The cultivation of grain goes much farther north than might be imagined. Barley ripens every year in Malangerfiord (69°), and even farther north, in Lyngen and Alten (70°); and is met with in the districts where Norway and Sweden join Russia. At Enontekis, 1350 feet above the sea, a little corn is grown, but it only ripens about once in three years. Thus the cultivation of corn is carried on in places where the mean temperature is below the freezing point, while in Switzerland it ceases at 9° above it, and in the S. American plateaux at $22\frac{1}{2}°$ above it; so that it is evident that the growth of grain is much more dependent on the summer temperature than on the annual mean. The long summer days of the polar regions afford a truly very brief but a comparatively exalted summer heat; but only barley attains such high latitudes. The following table exhibits the northern limits of the different grains :—

	On the west side.	On the east side.
	°	°
Barley	70	70
Rye	67	65—66
Oats	65	$63\frac{1}{2}$
Wheat	64	62

Wheat, however, is not cultivated to any extent beyond 60°. With regard to the altitude above the sea, barley goes to 800 feet in Southern Lapland (67°), while in Southern Norway (60°) it may be expected to ripen at an elevation of more than 2000 feet. The limits of the most important fruits are :—

On the west.   On the east.

Apples and plums $63\frac{1}{2}°$ $\left\{ \begin{array}{l} \text{(Tutteröe near} \\ \text{Drontheim)} \end{array} \right\}$ $62\frac{1}{2}°$ Sundsval.

Cherries........ 63   Ertvaagöe   63

Pears........... 62   62

Pears succeed commonly only as far as $64\frac{1}{2}°$ on the west, to $63°$ on the east. At Hammerfest in Norway $(71°)$, pears seldom ripen, while cabbage, turnips, carrots and spinach flourish. Potatoes also reach this latitude; but asparagus is only cultivated as far as $61°$ on the west side.

Southern Sweden, as already stated, is separated from the great Scandinavian Alps by a chain of large lakes little elevated above the sea, lying in a comparatively low tract of country. No large connected ranges of mountains are met with in this southern point, lying between $55°$ and $59°$ N.L., but several small mountains and ridges. The highest mountains are Taberg (1050 ft.) and Kinnekulle (830 ft.). The small ridges of Hallandsaas and Kullen do not rise so high, while the southern region Schoonen becomes perfectly level country. On the south and east lie the islands of Bornholm, Gothland and Œland; in the first occurs the hill of Rytterknœgten, the highest point being 480 ft.; in the others the heights are still more inconsiderable. Southern Sweden is rich in lakes, but poor in rivers: the mountains are composed partly of primitive, and partly of sedimentary rocks.

The temperature may be deduced from those of Stockholm and Lund, lying in the country of the south and north limits :—

	Lat.	Annual.	Winter.	Summer.
	°	°	°	°
Stockholm..	59½	42	25	62
Lund......	55½	45½	30	63

The mean heat in Lund, therefore, exceeds that of Stockholm by 3°, the cold of winter about 3½°, while the summer is only 1° warmer. The climate of Schoonen approximates more to the maritime character than that of Stockholm, as might be expected from their different positions.

The annual amount of rain amounts in Stockholm to 17½ inches, Westeras 17½ inches, Lund 18 inches, Wexio 21 inches.

The forests, like those of the Scandinavian mountains, consist chiefly of spruce and Scotch firs, and birch; but the beech is frequent in the southern parts. On the west coast this tree reaches Gottenburg (58°), on the east only to Calmar (56°, 57°). The oak occurs throughout, but most abundantly in the southern provinces. Schoonen, with its level surface and mild climate, is the most favoured spot in the Scandinavian peninsula for agricultural operations.

As Southern Sweden forms a south-eastern prolongation at the southern extremity of the peninsula, so Finland stretches out in the same direction at the northern point. Lying between 60° and 66° N.L., its breadth from north to south is about 90 geographical miles, and from east to west from 70 to 80. The two countries also agree in possessing low mountains and numerous lakes: the Œlandic islands form a natural bridge of connection between them. Finland may be regarded as a low mountain range, flat on the top, inclosing a great number of basins, or depressions, having lakes in their centres. The highlands fall off steeply towards the Gulf of Finland, but more gradually towards that of Bothnia. The greatest elevation scarcely exceeds 1200 feet. The number of lakes is more numerous even than in South Sweden, while the rivers are fewer. The highlands of Finland pass gradually on the one hand into the east flat part of the Scandinavian mountains, and on the other into the great east-European plain. The White Sea and Lake Ladoga form the most natural boundaries.

Like the Scandinavian peninsula, Finland is colder on the east than on the west side, the difference of the seasons being greater on the former: farther distant from the great ocean the annual mean temperature is still lower, and the differences in the different seasons still greater than on the east side of Scandinavia.

	Lat.	Annual.	Winter.	Summer.
	°	°	°	°
Umea	64	35	14	57
Uleaborg	65	33	10	58
Stockholm	$59\frac{1}{2}$	42	25	62
Abo	$60\frac{1}{2}$	40	23	60

The quantity of rain at Abo amounts to 20 inches: this is somewhat greater than in South Sweden, probably on account of the position between the two gulfs of the Baltic; but it does not nearly reach that of the west side of the Scandinavian Alps.

The forests are formed principally of Scotch fir and birch, but the oak is found as far as Biorneborg ($61\frac{1}{2}$°). The cultivation of grain is considerable.

The peculiar position of Scandinavia, connected as it is with the three great continents of the northern hemisphere, and favoured by the advance of the isothermal lines, causes it to present a richer and more varied flora than any other region situated so near the north pole. The western coasts, under the influence of the maritime climate, exhibit species which indicate the relation kept up, through the medium of Iceland and the Feroes, with Greenland and the other parts of Arctic America. The unbroken continuity with northern Asia has led to the presence of Siberian forms, flourishing under the somewhat continental climate existing on the east side of the Scandinavian Alps; while in Southern Scandinavia the milder climate permits the advance of the majority of the common plants of the North European flora.

Meeting on a comparatively confined space, and on a soil which exhibits the greatest variety of elevation of surface, it is natural that the limits of the different regions of vegetation should be rather indefinite in many cases: still, when broadly viewed, we obtain many striking contrasts and diversities. The flora of Lapland, the most northern part of the peninsula, has been very thoroughly investigated by the countrymen of Linnæus, who have followed the example of their great master, and devoted several special works to the description of the 'Flora Lapponica,' so attractive from its varied and relatively rich character. Entering therefore into more minute particulars than are involved in the above-mentioned general distribution of the arborescent vegetation and cultivated plants, we will in the first place examine the vegetation of Lapland.

The high ridge of the great mountain chain forms a natural division into two unequal parts, one of which, on the Norwegian side, falls rapidly towards the sea-shore; the other, larger sinks, as we have already stated, with more gradual descent; and we find the eastern boundary, where Swedish Lapland joins the narrow strip of Bothnia, skirting the western shore of the gulf of that name. The Swedish side is capable of subdivision into three regions of vegetation dependent upon elevation, and in these may again be distinguished several zones :—

1. Sylvatic or coni-
ferous region.
    a. Lower—mostly woody.
    b. Upper—mostly plains or low hills.
    c. Subsylvatic—mostly shores of lakes.

2. Subalpine or
birch region.
    a. Lower—*Betula glutinosa*, mostly rocky.
    b. Upper—dwarf birch (*B. nana*).

3. Alpine region,
destitute of trees.
    a. Alpine proper — drooping willow and bare plains.
    b. Snowy—plains, sometimes covered with snow.
    c. Glacial—summits of the Alps.

a. The lower sylvatic zone extends from the borders of Bothnia to a line drawn through Munioniska, Jockmock, Arfvidsjaur, Stor Afvonn, and Falträsk ; at which point the first bare hill-tops are met with : throughout it is distinguished by the abundance of vast swamps which extend in all directions among the woods, above which rise no naked hills, the few elevated spots being wholly overgrown with wood. Considerable rivers traverse this region, the largest often opening out into small lakes in their course. The soil of these tracts is mostly sandy, and the vegetation which flourishes in the gloomy shades and dismal swamps offers no striking feature, if we except perhaps the abundance of sedges. Most of the plants are related to more southern regions, and not more than twenty-two alpine plants descend into these uninviting localities. Many of the common wood and swamp plants of the general European flora occur here, and the following is a list of those which flourish throughout Lapland, up to the higher boundary of this zone :—

*Nymphæa alba, Calypso borealis, Gnaphalium sylva-
ticum, Mentha arvensis, Lysimachia vulgaris, L. thyr-
siflora, Ranunculus auricomus, Thalictrum rariflorum*
(Fr.), *Chelidonium majus, Lathyrus palustris, Polygonum lapathifolium, Salix aurita, Salix nigricans pru-
nifolia* (Lilj. Fr.), *Callitriche autumnalis, Calla palustris, Carex livida, C. acuta, Phleum pratense, Nardus
stricta.*

Many species of more southern character do not extend
all over this zone, while a certain number of alpine
plants descend into it from the subalpine region, such as
*Saussurea alpina, Cornus suecica, Aconitum Lycocto-
num, Tofieldia borealis, Astragalus alpinus, Rubus
arcticus,* &c.

b. In the upper sylvatic zone the forests are chiefly
composed of the spruce fir, but its upper limit is difficult of determination. Only in Tornea and Pitea does
the spruce form the boundary in certain places, in the
former between Vittangi or Ketkesuando, in the latter
between Lake Tjakelvas, Rappen, and Hornafvan: for
these parts of Lapland, especially the district of Tornea,
are very flat, having few mountains of any height, and
slope gradually and slowly towards the sea, Ketkesuando
lying at 1000 feet above it. Lulca and Umea, on the
contrary, present long and deep valleys up to the foot of
the high mountains which stretch out far and wide hither-
ward from the alpine ridge. The temperature is almost
incredible in the deep valleys, so that the spruce rises in
their heated soil much higher than the Scotch fir on the
slopes of the alps. In this part of Lapland, therefore,
the upper sylvatic zone should be defined to end where
the spruce ceases to grow in open tracts, as for instance
in Tjomotes, where it ascends to 1400 feet, and Gauträsk

near Naullofjellet. The upper part of this region thus lies at 1000 feet above the sea, or 3100 below the snow-line.

This region also is covered with dense forests and sterile marshes, so that its vegetation, which is of more southern origin, is altogether different from the preceding. Hills, too, lift their bare summits some 300 feet above the tree-limit, never forming any continuous ridges, but lying scattered, especially in Kemea, where Wahlenberg counted seventeen of them. On these mountains the true common alpine species begin to grow; while a vegetation, especially remarkable for the abundance of heaths and alpine willows, very different from that of the woods and swamps, covers the widely extended plains which stretch up the slopes of the mountains, where the woods are growing more and more rare. The following plants, common all over Lapland lower down, here attain their highest limit :—*Cirsium palustre, Hieracium rigidum, Galium Aparine infestum* (L. W. K.), *Limosella aquatica, Nasturtium palustre, Viola canina, Cerastium viscosum, Elatine hydropiper* (L. ?), *Chrysosplenium alternifolium, Polygonum amphibium, Rumex domesticus, aquaticus,* and *Carex leporina.*

Upon the mountains already mentioned, the following present themselves :—

From the subsylvatic zone : *Pinguicula villosa, Primula farinosa, scotica, Epilobium alpinum,* several *Salices, Carex heleonastes, Poa alpina,* &c.

From the subalpine region : *Gnaphalium norvegicum, Petasites frigida, Salix Lapponum, Phleum alpinum,* &c.

From the Alpine region : *Gnaphalium supinum, Diapensia lapponica, Primula stricta, Silene rupestris, Cerastium alpinum, Thalictrum alpinum, Saxifraga aizoides,*

Arctostaphylos alpina, Azalea procumbens, Menziesia cœrulea, Juncus trifidus, &c., &c.

c. The subsylvatic zone extends up as far as woods, chiefly composed of *Pinus sylvestris,* still exist ; and these form its most distinctive character. Its upper limit is again difficult to define, because, as we have already noticed, the spruce goes up the deep and narrow valleys beyond the pine, which is the last tree in the other parts of Lapland.

It varies in width from six or seven to two or three miles, being also interrupted in three different places by the existence of large lakes, at which it is altogether wanting ; and the alpine region immediately succeeds the sylvatic. The tolerably dense woods which occur here are of lower stature than below, indicating the influence of the neighbouring alps ; and diminishing gradually under the increasing power of frost and cold, they become few, dwarfed and gnarled, scattered about the marshes and between the cliffs till they wholly succumb to the hard climate. The swamps are less numerous than in the preceding regions, but the largest lakes of Lapland lie within the boundaries of this zone. Here, therefore, commences the so-called littoral vegetation, remarkable for the abundance of willows, decking the banks with their thickly interwoven bushes, as well as of most of the aquatic plants of Lapland. Here reside the hardiest of the husbandmen, tilling fields which often deny a harvest to the sickle, and cultivating barley and potatoes in a few open patches. This zone presents nothing remarkably distinctive in its vegetation. The following southern plants, common to the two lower zones, ascend to the upper boundary throughout :—

Pyrethrum inodorum, Galium boreale, Myosotis pa-

*lustris, Veronica scutellata, Prunella vulgaris, Plantago
lanceolata, Nuphar intermedium* (Ledeb.), *Ranunculus
flammula, Serapis arvensis, Turritis glabra, Oxalis aceto-
sella, Drosera rotundifolia, longifolia,* and *intermedia,
Rubus Idæus, Potentilla anserina* and *tormentilla,
Saxifraga Hirculus, Trifolium pratense, Pyrola uniflora,
rotundifolia,* and *secunda, Goodyera repens, Maian-
themum bifolium, Scheuchzera palustris, Potamogeton
gramineus,* and *prælongus, Pinus sylvestris, Abies excelsa,
Juniperus communis,* &c.

From the subalpine region descend, *Myosotis alpestris, Veronica alpina,* and, in places, *Erigeron elongatum*
(Ledeb.), *Luzula parviflora* (Ehrh.), and *Sonchus sibiricus.*

From the Alpine region, *Archangelica officinalis*
(Hoffm.), *Viola biflora, Oxyria digyna, Salix myrsi-
nites, Calamagrostis phragmitioides* (Hn.), and *Agrostis
rubra* (Whbg.).

II. The subalpine region, or that of the birches, might
be divided into two zones; a *lower,* containing the bases
and sides of the alps, remarkable for the bright groves
of birch (*Betula glutinosa*); and an *upper,* comprehending the upper flanks and the commencement of the
mountain plains, where the dwarf birch (*Betula nana*)
alone occurs with its low and creeping stems. But it is
impossible to carry out this distinction, since the *Betula
glutinosa,* which is of sufficiently lofty stature at the foot
of the alps, becomes more and more dwarfed as it ascends;
and thus where it occurs above intermixed with the dwarf
birch, it becomes exceedingly like it: moreover, the
plants which principally inhabit this region appear to
occur without distinction in all parts of it.

The lower limit of this region is not easily defined, on

account of what has been said respecting the upper limit of the pine ; and its upper boundary is still more undefined. Where the birch is seen clothing the sides of the mountains, it is marked clearly enough, near Quickjock ascending to 2200 feet above the sea and 360 above the limit of the pine. The birch there forms groves ; quickly becoming stunted to some six feet high, shrubby, and at length ceasing altogether. Most frequently, however, this region penetrates into the valleys bordering them, from the foot of the mountains up to the alpine ridge, so that the birch sometimes grows in abundance in the valley of the mountains on the north side, while the exposed southern foot is covered by snow and ice : it ascends especially high on the slopes which bound these valleys on the side next the great alpine ridge. In those places where great depressions occur, this region even crosses the ridge, connecting the birch groves of Swedish Lapland with those of Nordland and Finmark. The general upper limit, however, may be stated at 2100 feet above the sea, which is also about 2000 feet below the snow line. The vegetation of this region is very different from that of the three preceding zones, not only in aspect and nature, but in the abundance of certain species and forms. Very varied stations occur here, causing great diversities in the plants. The rivers and brooks descending from the alps display aquatic plants, such as *Hippuris*, *Myriophyllum* and *Potamogeton*, all remarkable for great elongation, flaccidity and variation of form, growing as they do in water running with great velocity in summer, and often wholly frozen up in winter. In the woods which grace the sloping sides of the mountains is found a vegetation which is seldom equalled in Scandinavia for magnitude and luxuriance, produced here

by the extreme heat of the short summer, together with
the plentiful supply of streamlets watering it, and the
abundance of humus, which is brought down by violent
tempests, the winds it is said carrying along particles of
organic matter with them. This holds good most particularly of the rugged sides of the mountains, where the
southern vegetation, mixed intimately with the northern,
flourishes in a remarkable manner ; for the heat favours
the ascent of the southern species to these points, while
the brooks and the stones rolling down bring the alpine
plants from the higher plains. Thus whether we consider the lower part where the southern vegetation principally flourishes, or the upper where the alpine plants
come down most plentifully, this region is most rich, and
pre-eminently exhibits those characters, for which an alpine region is most remarkable.

The following is a list of some of the southern plants
which are common to the lower regions, and up to the
superior boundary of this :—

*Achillea millefolia, Carduus crispus, Taraxacum offi-
cinale, Valeriana officinalis, Asperugo procumbens, Me-
nyanthes trifoliata, Veronica officinalis, Rhinanthus
minor* (Ehrh.), *Carum carui, Brassica campestris, Barbarea stricta, Geranium sylvaticum, Silene inflata,
Stellaria nemorum, Spergula arvensis, Montia fontana,
Epilobium palustre, Myriophyllum alterniflorum, Hippuris vulgaris, Comarum palustre, Spiræa Ulmaria,
Cerasus Padus, Polygonum Convolvulus, Urtica urens,
Betula glutinosa, Alnus barbata* (C. A. M.), *Callitriche
verna, Orchis maculata, Potamogeton natans, rufescens,
perfoliatus, pusillus,* numerous *Carices, Aira flexuosa,
Milium effusum, Alopecurus geniculatus,* and *fulvus,* &c.

From the alpine region descend :

Gnaphalium alpinum, Erigeron uniflorum, Hieracium alpinum, Veronica saxatilis, Gentiana nivalis, Arabis alpina, Saxifraga nivalis, stellaris, oppositifolia, cæspitosa and *adscendens, Alchemilla alpina, Sibbaldia procumbens, Dryas octopetala, Andromeda hypnoides, Salix reticulata,* &c. &c.

III. The Alpine region, containing the mountain summits devoid of trees, comprehends three zones :—

A. From the alpine ridge which traverses Scandinavia, dividing Norway from Sweden, isolated mountains pass off into the various regions, which are indeed for the most part inhabited by truly alpine species, such as grow upon the peaks and plateaux, but since the last of the southern plants (a very few only excepted which go even higher) occur in these alps in places which are free from snow-heaps throughout the summer, a sub-region, the proper Alpine zone, may be distinguished. Such isolated alps occur in Umea, Pitea and Lulea, but not in Tornea.

The number of species and individuals diminishes in proportion to the altitude, and the zones which may be distinguished upon mountains are so much the more broadly marked when the alps rise to a great height. In Lapland, however, where the summits are comparatively little elevated above the plains, the discrimination becomes very difficult, since the different regions become crowded into a small space ; for while the Alpine region of central Europe commences at 6000 feet, that of Lapland is entered at 2000 feet.

The last Birches disappearing, large plateaux extend in all directions, the earlier portions of their gradual slopes being clothed with willows, chiefly *S. glauca, lanata* and *Lapponum* ; under the shade of which cower many plants of southern origin, which, common in all the

H

lower region, attain their limit in this, such as : *Cirsium heterophyllum, Leontodon autumnale, Pinguicula vulgaris, Ranunculus acris, Trollius europæus, Habenaria viridis, Listera cordata, Andromeda polifolia, Carex vesicaria* and *ampullacea*, &c. &c.

The bare mountain plains to which the Willows do not extend, flat or rising again slightly beyond, are covered with a true alpine vegetation, with which very few southern plants are intermingled ; heaths forming dense tufts, rushes collected in the dampish places, and sedges filling the small but numerous swamps, form the most striking features.   The snow all melts in June, and except in very cold summers, snow-heaps very rarely remain even in the more obscure places.   From the marshes formed by the water stagnating in the hollows, innumerable rivulets flow down, irrigating the bare plains, and, taking their course through the willow beds, reach the region of the birch, collecting the most beautiful of the alpine plants upon their banks.   Besides the alpine species already mentioned as descending to the lower regions, the following occur throughout this zone :—

Ranunculus nivalis, pygmæus, Silene acaulis, Cerastium trigynum, Saxifraga comosa (Poir.), *Potentilla alpestris geranioides, Salix glauca-pullata, glauca-pallida, herbacea, Carex vesicaria-alpigena, C. saxatilis, ampullacea-borealis, ustulata, hyperborea, lagopina, rupestris, Poa cæsia, Aira alpina, Agrostis rupestris.* And in some places: *Saxifraga cernua, Erigeron alpinum, Braya alpina, Poa cenisea,* with its variety *flexuosa,* and *Phaca frigida.*   The plants intermixed with these alpine species (most frequent in the lower regions) are *Campanula rotundifolia, Trientalis europæa, Rubus Chamæmorus, Vaccinium Myrtillus, Empetrum nigrum, Polygo-*

*num viviparum, Betula nana, Gymnadenia Conopsea,
Luzula campestris* and *nivalis, Carex cæspitosa, aquatilis, vulgaris* and *chordorhiza, Hierochloe borealis,* and
Anthoxanthum odoratum.

The limit of these plains is formed by those higher
summits which, covered for the most part by snow-fields,
only produce on their lowest parts the plants inhabiting
the higher alps. Besides *Silene acaulis, Alsine biflora,
Ranunculus nivalis* and *pygmæus, Cerastium trigynum,
Saxifraga cernua, Salix herbacea* and *Poa cenisea,* which
are in greatest luxuriance here, but also descend lower,
there are found also :

*Ranunculus glacialis, Cardamine bellidifolia, Draba
nivalis, Salix ovata, Luzula glabrata, arcuata,* and *Poa
laxa.*

B. Plains spreading among the higher alps, where the
snow lies everywhere throughout the year.

Beyond those alps which stretch out from the main
ridge lie vast plateaux, formed by the gently sloping
sides of the ridge, and most frequently occupying the
intervals between the alpine peaks. Here no tree flourishes, not even the dwarf birch ; in places exposed to
the sun only, and where the soil is especially favourable,
willows (*Salix hastata-alpina*) emerge with short and
twisted branches ; *Salix polaris,* however, is most frequent among the rocks, and with *Salix herbacea* constitutes the entire arborescent vegetation. The snow never
entirely disappears from this region ; those places even
which are completely uncovered by snow, are continually
and abundantly overflowed by rivulets, coming down from
the glaciers above. Thus that zone may be called the
snowy, which includes these sloping or level plains nearest
to the alpine ridge and where the snow never melts en-

tirely, and which often rises to within 200 feet of the snow-line. The boundary cannot be said to extend to the snow-line, because few of the summits rise so high, and when they do they are covered by vast glaciers descending from below this line into the snowy zone, so that all vegetation is arrested.

The plains among the Alps are everywhere inhabited by *Pinguicula alpina, Saxifraga rivularis,* and *Carex microglochin.*

The following species are more local :—*Arnica alpina, Campanula uniflora, Pedicularis flammea* and *hirsuta, Draba alpina, Oxytropis lapponica, Andromeda tetragona, Salix arbuscula, Gnaphalium carpaticum, Wahlbergella apetala, Salix polaris, Carex rariflora, Cerastium latifolium, Kœnigia islandica, Carex bicolor, fuliginosa* and *limula.*

On the sloping and rocky sides of the alps, below the snow-line, flourishing together with *Ranunculus glacialis, nivalis, pygmæus, Dryas octopetala, Cardamine bellidifolia, Draba nivalis* and *Luzula arcuata,* which ascend exceedingly high, occur locally :—

Alsine rubella, Alsine stricta, Arenaria humifusa, Salix Myrsinitis-procumbens, Betula alpestris (Fr.), *Luzula hyperborea, Carex nardina, Rhododendron lapponicum, Gentiana tenella, Kobresia Scirpina, Salix arbuscula-vaccinifolia, Chamæorchis alpina* (Rich.), and *Catabrosa algida.*

C. The alpine ridge and its peaks are for the most part covered with snow and ice. The ridge extends from 32° to 39° E. Long., and from 64° to 69° N.L., not uninterruptedly, but broken in three places, as above mentioned, within the limits of Lapland. From the highest peaks vast sloping fields of snow stretch down into the lower regions, the lower part being there gradually melted by the heated surface which they invade, but the waters which flow down from above are frozen in the cold nights, forming glaciers. They frequently descend into deep valleys, but their edges gene-rally correspond to the snow-line. Hence, when their lower parts have descended far, they heap up before them walls of humus, *moraines*, which, watered by the melting snow, do not form a solid and hard mass, but are black, soft and turgescent, and upon them occur the first (or last) plants, such as *Ranunculus glacialis, Saxifraga cernua* and *glacialis*, &c., and the intermingled rocky fragments become clothed with Lichens.

The snow-line, which in the alps most distant from the ridge, as in Wallivara near Quickjock, attains a height of 4100 feet, falls in these spots where all around is cold and heaps of snow abound, to a height of only 3100 feet above the sea, the mean temperature of the region being about two degrees above the freezing-point.

Western, or Norwegian Lapland.—The region lying on the Atlantic side of the mountains includes the provinces of Nordland and Finmark, and is of very small breadth, sometimes only amounting to 4–6 geographical miles, forming as it were only a narrow zone between the alpine ridge and the sea. Its shores are washed by the waves of the western ocean and the Icy Sea, and are indented by deep and narrow bays, like the valleys which occur on the Swedish side, these bays being surrounded by lofty and precipitous mountains gradually rising towards the ridge. Hence it is difficult to make a distinction into natural regions of vegetation, but on the other hand the differences of kind and frequency of alpine plants depending on altitude and climate are seen almost in one view.

This part of Lapland, when compared with the Swedish side, displays much that is different from that and peculiar to itself. In the maritime climate which this side possesses in common with the island regions lying round the pole, it agrees closely with Norway, in the same way as the other part of Lapland possesses, as already indicated, great resemblance to Siberia. Thus we find here a maritime region, altogether wanting on the Swedish side, inhabited by plants above all peculiar to the Norwegian side.

As the vicinity of the sea produces greater equality and mildness of temperature than exists on the Swedish side, and as this temperature differs according to the latitude, it is possible to determine more accurately here in what places the plants cease towards the north. The great variety of temperature is shown by the following figures :—

	Nidarosia.	Island of Mageroë.
Annual mean	40° Fahr.	32°
Mean of January	$19\frac{1}{2}$	22
Mean of July	$64\frac{1}{2}$	40

It is not wonderful therefore that many plants of more southern character occur here which have not been met with on the Swedish side.

As far as Loffoden are found *Ranunculus polyanthus, Hypericum quadrangulare, Trifolium medium,* &c.

The following proceed as far as Nordland :—*Knautia arvensis, Stachys sylvatica* and *palustris, Origanum vulgare, Linum catharticum, Mœhringia trinervis, Saxifraga tridactylites, Rosa canina, Vicia sylvatica, Corylus Avellana, Listera ovata, Stratiotes aloides,* &c.

As far as Northern Finmark :—*Glechoma hederacea, Gentiana campestris, Corydalis fabacea, Lychnis flos-Cuculi, Vicia sepium, Lathyrus pratensis, Convallaria verticillata,* &c. &c.

The following occur, scattered, as far as the river Alten:—*Hieracium Lawsoni, Veronica Chamædrys, Viola palustris, Lythrum Salicaria, Herminium Monorchis, Epipactis latifolia.*

And as far as the river Tanea and Varangefjord :—*Thymus serpyllum, Sedum acre, Sonchus arvensis* and others.

Many even go on further to the north ; but the following, which occur only in the most southern parts of the Swedish side, advance almost to the North Cape :—*Veronica officinalis, Ajuga pyramidalis, Vicia Cracca, Crepis tectorum, Artemisia vulgaris, Erigeron acre, Senecio vulgaris.*

It is evident from these notices that Norway is far more rich in plants, and this is further shown by the fact that only twelve species of the Swedish side are at present deficient in the lists of the Norwegian.

The regions which may be distinguished on the West side of the Scandinavian Alps are :—

1. The Maritime region ;
2. The Subsylvatic region ;
3. The Subalpine region ; and
4. The Alpine region.

All excepting the maritime agree to a great extent with the corresponding regions on the Swedish side, so that it is unnecessary to quote species common to both ;

in the following remarks the regions are briefly characterized, and only their *peculiar* species noted.

A. *The Maritime region.*—This region comprehends the Norwegian shore up to the most northern promontory; it does not form a continuous zone, since the bays or fiords frequently interrupt it. The sea-breezes nourish many plants peculiar to this region, such as :—*Aster Tripolium, Steenhammera maritima, Plantago maritima, Glaux maritima, Armeria elongata* (Hoffm.), *Cakile maritima,* &c., together with *Primula finmarkica, Gentiana involucrata, Allium sibiricum, Carex norvegica,* and many others.

B. *The Subsylvatic region.*—Containing spots occupied by Scotch fir.

Although, as has been said, many southern plants advance much further north on the Norwegian side, the spruce (*Abies*) only goes to 67°, not nearly so far as on the Swedish side, no doubt on account of the unfavourable influence of the sea-air. The Scotch fir, which bears the maritime climate well, goes to 70°; to this degree it forms either a continuous zone, or small forests detached from each other in those deep and narrow valleys which occur especially in the higher mountains of Nordland, flourishing most vigorously in Saltea and Tromsoë. More to the north it fills only the interior bays which penetrate as far as the alps; as in Western Finmark, and in the district of Fælles, it forms but a narrow zone, ascending 368 feet on the eastern inclines of the mountains. The valleys just mentioned present a luxuriant vegetation of those more southern plants which we have described as advancing towards the north, together with others almost the same as in the corresponding region of

Swedish Lapland, while a few are peculiar to this part : *Erigeron rigidum* and *politum* (Fr.), *Conioselinum tataricum*, *Papaver nudicaule*, *Draba incana*, *Oxytropis campestris*, *Salix finmarkica*, and *Platanthera obtusata*.

C. *Subalpine region or region of Birches.*—As on the other side, the region of the firs is preceded by a region of birches. In the province of Alten the birch-woods reach 1000 feet, and in the deep valleys, as at Sorrefiord, as far as 1479 feet; but when they penetrate further towards the alps the Birches cease sooner, so that on the alps of Sulitelma, on the lake Sorjasjaur, the northern, exposed faces are clothed with brilliantly green woods, while the southern slopes have none. Towards the north this limit becomes a little depressed, so that the birch does not ascend higher than 1200 feet in Southern Finmark, and it is said grows only to 725 at Hammerfest ; then in the latitude of 70°, where the Scotch fir disappears, descending from 368 towards the sea-shore, it finally disappears at the North Cape. In this region almost all the subalpine species occur commonly, such as :—*Archangelica*, *Viola maritima*, *Epilobia*, *Sonchus alpinus* and *sibiricus*, *Menziesia cærulea*, *Salices*, *Carex atrata*, &c. But when the birch descends to the shore we find these plants, especially the succulent, such as *Rhodiola rosea*, *Primula stricta*, *Draba*, &c., occurring near the sea, and almost become maritime. Besides those already noted as common throughout Scandinavia or Lapland, not a few occur peculiar to this part; as :—*Arenaria lateriflora*, *Chrysosplenium alternifolium*, *Salix punctata*, *Salix herbacea-fruticosa* (Fr.), *Veratrum album* (*Lobelianum*), L., *Triticum violaceum*, *Catabrosa latifolia*, &c.

D. *Alpine region.*—This is the most extensive of all.

The ridge exhibits more peaks and valleys on the Norwegian side than on the Swedish, the highest of which
are situated around the town of Tromsoe and the Lyngenfjord. Penetrating through Western Finmark the
alpine region bends nearer to the Icy Sea, and becomes
gradually lower till the mountains of Worieduder and
Rastekavic, scarcely attaining the snow-line, form its
boundary. Since the climate of this side is maritime,
the clouds are collected against the ridge, and water the
soil with abundant rains, which in the summer season
extensively thaw the abundant snow which falls in water,
and then, the water freezing, innumerable glaciers are
produced. In Nordland they are most numerous from
Kunnen to Lyngenfjord (about twenty are found there) ;
in Finmark they occur only on three islands around
Hammerfest. These fields of snow and ice, covering so
large a portion of the alpine ridge, destroy all vegetation
there, and it is not surprising that the plants which
flourish best among the melting snows are the succulent,
such as *Rhodiola, Ranunculus nivalis* and *glacialis* ;
that the alpine shrubs, such as the heaths and *Diapensia*, are neither frequent nor abundant in this part of
Norway, and that the true alpine plants must be sought
chiefly on the lower parts of the flanks. Besides the
plants mentioned as existing on the Swedish side, the
following have been gathered here :— *Wahlbergella
affinis, Saxifraga Aizoon, Sedum villosum, Kobresia
caricina,* &c.

With regard to the region formed by the willows, this
is one of the most marked in Norway. Not only *Salix
lanata* and *glauca,* but also *ovata* and *Myrsinitis* contribute to form it, and those plants are also met with which

have already been stated to flourish best among these. This region rises to 1080 feet at the river Tanea, near the island of Mageroë only to 470.

In addition to these regions must be noticed the maritime alps detached from the main ridge, and most frequently precipitous, forming as it were separate ridges, the peaks of which, sometimes covered with ice, mostly surpass the snow-limit. To this class belong not only those mountains, the broken and steep sides of which form the coast of Norway, consisting of arms projecting from the alpine chain, but also those constituting the mountainous islands, which form the archipelago studding the shores. Exposed to the most violent blasts of the sea-winds, they nourish neither trees nor scarcely shrubs, so that the junipers and andromedas are seen rarely there ; the saxifrages, however (chiefly *S. oppositifolia*), and succulent plants are luxuriant. Two plants are peculiar : *Ranunculus sulphureus* and *Saxifraga hypnoides*.

Russian Lapland.—This part of Lapland has not been so thoroughly investigated as the other regions, but so far as is known it contains many Siberian plants not as yet met with in them. The alpine ridge does not extend into it, but is lost by degrees, succeeded at first by detached mountains, till the wide peninsula between the Varangerfiord and the White Sea appears flat and filled with lakes, rivers and morasses. Here it is very difficult to determine regions of vegetation, and the plants of southern and of arctic origin appear in some degree intermixed. Thus, southern plants occur as far as the town of Kola, which have not been found either in Swedish or Norwegian Lapland :—

Ribes nigrum, Cotoneaster vulgaris, Orobus vernus,

Heracleum sibiricum, Salix amygdalina, Gnaphalium uliginosum, Schedonorus inermis.

And although that town is situated on the most northern shore of the peninsula, *Viola tricolor, canina* and *palustris, Silene nutans, Sedum acre, Rubus Idæus, Fragaria vesca, Lathyrus pratensis,* and many more are common, which are only found in the southern parts of the other Lapland provinces.

Finally, some twenty-nine species appear peculiar to this region, among which are : *Chrysanthemum arcticum, Pyrethrum bipinnatum, Ligularia sibirica, Aster sibiricus, Cenolophium sibiricum, Rosa carelica* (Fr.), *Hedysarum obscurum, Carex halophyla,* &c. &c.

Altogether the Lapland flora comprehends about 685 species, containing 283 genera and 72 families, which may be referred to three classes :—1. Plants truly alpine, occurring also in the other alps of Europe, as in Scotland and Switzerland ; these amount to about 108. 2. Plants truly arctic, which have not yet been found except in northern countries, and may be regarded as peculiar to them ; there are 97 in Lapland proper and 27 in Russian Lapland, making together 124 arctic plants.   And 3. Southern plants common not only to the other parts of Scandinavia but also throughout Europe, comprehending the remaining 453 species.

The families which contain the greatest number of species are : Cyperaceæ $\frac{1}{8}$ of the whole flora, Compositæ $\frac{1}{10}$, Grasses almost $\frac{1}{10}$, Cruciferæ $\frac{1}{21}$, Salicineæ $\frac{1}{23}$, Ranunculaceæ $\frac{1}{27}$, Alsinaceæ $\frac{1}{27}$, Juncaceæ $\frac{1}{27}$, Rosaceæ $\frac{1}{23}$, Scrophularineæ $\frac{1}{32}$, Leguminosæ, Ericaceæ and Orchideæ $\frac{1}{36}$, Saxifrageæ $\frac{1}{48}$, Labiatæ and Umbelliferæ $\frac{1}{52}$, and Caryophyllaceæ $\frac{1}{17}$.

Of the 697 species, 469 are Dicotyledons and 228 Monocotyledons.

The following families occurring in the general flora of Scandinavia (including Denmark) are not represented in Lapland : Ambrosiaceæ, Loranthaceæ, Lobeliaceæ, Convolvulaceæ, Oleaceæ, Verbenaceæ, Asclepiadaceæ, Solanaceæ, Globularieæ, Araliaceæ, Aceraceæ, Resedaceæ, Berberaceæ, Balsamineæ, Tiliaceæ, Malvaceæ, Cistaceæ, Cucurbitaceæ, Paronychieæ, Aristolochiæ, Santalaceæ, Eleagneæ, Ulmaceæ, Amaranthaceæ, Ceratophylleæ, Iridaceæ, and Narcissineæ.

The vegetation of Lapland has been treated at great length, since it possesses so peculiar a character that it affords an excellent representation of the arctic conditions, and a somewhat definite acquaintance with it will enable the reader to comprehend more clearly the contrast between the northern and southern regions of Europe. The remainder of the Scandinavian peninsula must be spoken of more briefly.

With regard to that part of Norway extending south from Nordland, the country widening out still retains its mountainous character, and the plateaux upon this become much more extensive, while the maritime climate is equally visible on the sea-coast, and the advance towards the south is marked by the successive appearance of many more southern species occurring in the other districts of the European flora. A comparison of the numbers yielded by West Finmark and the entire Norwegian flora shows that out of 84 families possessed by the latter, comprehending 404 genera and 1200 species, West Finmark has only 50 families, 177 genera, and 402 species.

But the 34 families more occurring in Southern Norway are very feebly represented, 20 of them having but a single genus, and 13 of them only one species, showing that they are evidently of a southern character.

The following statistics of the entire Norwegian flora will indicate its relations to that of Western Lapland:—

The Ranunculaceæ form $\frac{1}{30}$, Cruciferæ $\frac{1}{21}$, Caryophyllaceæ $\frac{1}{21}$, Leguminosæ $\frac{1}{25}$, Rosaceæ $\frac{1}{18}$, Umbelliferæ $\frac{1}{32}$, Compositæ $\frac{1}{10}$, Ericaceæ $\frac{1}{52}$, Boragineæ $\frac{1}{57}$, Scrophularineæ $\frac{1}{28}$, Labiatæ $\frac{1}{33}$, Polygoneæ $\frac{1}{57}$, Salicineæ $\frac{1}{36}$, Orchidaceæ $\frac{1}{48}$, Junceæ $\frac{1}{46}$, Cyperaceæ, $\frac{1}{11\frac{1}{4}}$, Graminaceæ $\frac{1}{12\frac{1}{2}}$.

Of the Natural Families occurring in the Scandinavian flora, the following are not represented in Norway: Ambrosiaceæ, Verbenaceæ, Globulariaceæ, Cucurbitaceæ, Aristolochiaceæ, Santalaceæ, Ulmaceæ, and Amaranthaceæ.

Among the plants appearing in Norway alone are: *Erica cinerea, Rosa alpina, Saxifraga hieracifolia, Geranium macrorhizum, Bunium flexuosum, Gentiana purpurea, Campanula barbata*; the generality of its plants, however, extend either into Western Lapland on the north, or into Sweden on the south-east and south.

The following list contains plants only occurring in Norway on the western coast, not in the inner districts, but occasionally in Sweden; the Swedish locality is mentioned in a parenthesis, and the figures denote the extreme N. L. in Norway:—

Fumaria capreolata 59°, *Hypericum pulchrum* 63½°, *Vicia Orobus* 62½°, the limit of the oak, *Sanguisorba officinalis* 60° (also in the I. of Gothland), *Ilex Aquifolium* 62½° (also in Bohuslan), *Galium saxatile* 62½°

(south of Sweden), *Erica cinerea* 62½°, *Lysimachia nemorum* 63° (Schonen), *Digitalis purpurea* 63° (Bohuslan), *Lamium intermedium* 61°, *Teucrium Scorodonia* 59°, &c. &c.

The next list contains plants of the west coast which also occur on the south coast, but not in the inner districts :—

Arabis petræa 62°, *Pyrus Aria* 63½° (also in Bohuslan), *Hedera Helix* 60½°, *Lonicera Periclymenum* (Bohuslan), *Sambucus nigra* (Bohuslan), *Fagus sylvatica* 61° (Bohuslan), *Quercus Robur* 62°, or according to Blom 63° (south of Sweden), &c.

Certain inland plants of the eastern districts are absent on the western coast, such as :—

Pulsatilla vernalis, Trollius europæus, Berberis vulgaris, Astragalus glycyphyllos, Pyrola chlorantha, &c. &c.

The following plants of the eastern plateaux have been observed principally in the Dovrefjeld, but not on the west coast :—

Ranunculus hyperboreus, Lychnis apetala, Oxytropis lapponica, Phaca oroboides and *frigida, Saxifraga cernua, Primula stricta, Kœnigia islandica, Juncus arcticus, Kobresia caricina,* &c.

Thus the alpine vegetation of the Norwegian highlands seems to attain its maximum on the Dovre Mountains, in respect to the number of species, diminishing to the west and south. The individual numbers of many characteristic species are also said to decrease, the great plateaux gradually assuming the condition of a steppe. The predominating plants of such tracts are often *Molinia cærulea* and *Solidago virgaurea,* growing so vigorously as to displace all others. The alpine plants of this region resemble those of the more northern mountains and those of Scotland, but differ much from those of the Brocken and the Riesengebirge of Germany.

Under the head of Sweden we include all that portion of the peninsula south of Swedish Lapland (including the narrow strip of coast called West Bothnia), bounded on the west by the alpine chain, and a lake extending southward to the Cattegat which bounds South Sweden on the same side ; on the east by the Gulf of Bothnia and the Baltic, which also washes the southern coasts.

All that portion lying to the north of the chain of lakes forming the boundary of Southern Sweden partakes to a certain extent of the northern character. The alps extending continuously through Jemtland and Herjedal terminate, as in the north, with isolated mountains, in Dalecarlia. The same vigour of vegetation may be traced here as in Lapland, but it differs in relative development. The alpine regions extend down, decreasing in importance as far as Dalecarlia, and are succeeded as above by the *sylvatic lacustrine* region, including Jemtland, Herjedal, and much of Dalecarlia. The lakes originate through the alpine streams being dammed up by the low wooded, rocky or sandy hills, which are interspersed between these regions and the eastern littoral tracts. They differ in breadth in different places ; in Jemtland they are very wide, and in Herjedal again narrowed.

The lower, more maritime regions are situated along the coast of the Gulf of Bothnia, and differ in width according to the irregularities of the sylvatic regions above, being interrupted where the hilly country advances towards the sea ; thus West Bothnia is level and alluvial, while Medelpadia is a sort of continuation of the highland forming the lacustrine boundary of Jemtland. And other

differences exist between West Bothnia and Norrland
properly so-called, as well as exceptions in the other regions; thus the great river of Angermannia flows through
the smaller lakes and produces fluviatile islands characterized by *Tamarix germanica* and *Salix amygdalina*; in
Helsingen also a sort of lacustrine region exists by the sea-
shore around the lakes of Dellen, &c. Dalecarlia forms
a sort of delta between the rivers flowing into the Gulf of
Bothnia and Lake Malar, and is more related to S. Sweden
than the provinces above.

The following sketch will afford an idea of the gradual
disappearance of southern species towards the north.

Anemone Pulsatilla is scarcely seen beyond Upsal.
*Quercus Robur, Rhamnus catharticus, Cratægus Oxy-
acantha, Prunus spinosa, Centaurea Scabiosa, Ornithoga-
lum luteum, Turritis glabra*, &c., end at the river Dal-
elfven. In the neighbourhood of the town of Gefle end,
*Saxifraga granulata, Laserpitium latifolium, Malva ro-
tundifolia, Typha angustifolia, Carex intermedia*, and
others; while *Alnus incana* is found more plentifully,
and at Trodje *Betula nana* begins to grow copiously.
The borders of northern Gestricia form the northern
boundary of *Spiræa Filipendula, Trifolium montanum,
Primula veris* (sometimes seen sparingly at Sundeval),
Viburnum Opulus, Lythrum Salicaria (which however occurs on the sea-shores as far as Vesterbottnia), and pretty
nearly of the hazel; while the proper Lappic plants,
Rubus arcticus and *Carex globularis*, become very plentiful. About Hudiksval in Helsingen *Cornus suecica* is
very abundant; and here more inland, *Alnus glutinosa*,
as well as *Scabiosa succisa, Lychnis dioica, Carex erice-
torum*, and *Lemna minor* fail. In Medelpadia *Aconitum*

I

Lycoctonum becomes common, while the northern boundaries of *Knautia arvensis, Lonicera Xylosteum, Plantago media, Cynosurus cæruleus, Festuca fluitans, Arctium Lappa, Campanula persicifolia,* and *Galium verum,* are recognized. Beyond the remarkable Wood of Skulskogen, or in southern Angermannia, *Alnus glutinosa,* called by the natives the " Sea Alder," grows no longer even on the sea-shores, and at the same time *Acer platanoides, Anemone Hepatica, Rosa canina* (which also occurs on the sea-shore alone in Norrland), *Heracleum Sphondylium, Dactylis glomerata, Orobus tuberosus, Thymus Serpyllum, Lychnis Viscaria,* &c., cease. In north Angermannia we at length miss *Anemone nemorosa, Veronica Chamædrys, Pimpinella Saxifraga, Pteris aquilina,* and *Hypericum perforatum,* but *Salix limosa* occurs in great abundance. Finally, approaching Lapland, in Vesterbottnia, near Umea, we see for the last time, *Centaurea Cyanus, Salix fusca, Potentilla argentea* and *anserina, Dianthus deltoides, Lycopsis arvensis* and *Spergularia rubra,* and many plants which are only found on the sea-shores so far north, *e. g. Lythrum Salicaria* and *Sedum Telephium.* On the other hand, *Phleum alpinum, Salix arbuscula,* &c., present themselves. These limitations are of course general, the plants being said to end where they cease to grow in any plenty.

Sweden Proper includes that portion of the country lying around the great lakes separating S. Sweden from the rest of the Peninsula. The most eastern part forming a portion of the delta of the river Dale, is for the most part flat and fruitful ; its vegetation has a very eastern character, exhibiting *Salvia pratensis, Fumaria parviflora, Scutellaria hastifolia, Lavatera thuringiaca* and other species belonging to the opposite coasts of the Baltic. Some plants of the open country have their northern limit here.

The remaining portions of this district which have their watershed either into Lake Malar or Lake Hjelmar, agree partly with the eastern coast and partly with the adjoining tracts of Dalecarlia.

In S. Sweden may be distinguished three regions, under the names of East Gothland, West Gothland, and South Gothland.

1. East Gothland, including the islands of Œland and Gothland, exhibits a great diversity of character in proceeding from the east to the west. About Westervik, on the coast, the vegetation is luxuriant, and among other plants occur *Orchis sambucina, Corydalis cava,* and even *Trapa natans* ; more to the south the plains are not so rich, yet not very different ; but in ascending towards the interior, *Cynanche Vincetoxicum, Lathræa, Sanicula, Lonicera Xylosteum, Rhamnus catharticus,* and others are lost, and at the highest part are met *Thesium alpinum* and *Potentilla acaulis.* About Helgaijo *Melampyrum nemorosum, Pulmonariæ, Viola mirabilis* and *hirta,* and others are lost, and *Erica Tetralix* is met with. Next are lost *Spiræa Filipendula, Saxifraga granulata, Helianthemum vulgare,* &c.; and on the west, past the Laga, nothing of the luxuriant vegetation of the east seems to remain, while the beech, *Gentiana Pneumonanthe, Hypochæris,* &c., give a more autumnal character to the vegetation. Then *Narthecium ossifragum, Convallaria verticillata, Habenaria albida,* &c., appear as we advance nearer to the western coast.

2. West Gothland. The northern part, continuous with Dalecarlia, agrees in some measure with it, but

Radiola and *Gentiana Pneumonanthe* give it a western character, while Lake Venner is surrounded by *Erica Tetralix* ; *Genista tinctoria* is also characteristic of this province, while on the coast, although exposed so much to the sea, pine forests exist, and many plants which belong more properly to the more southern part of the coast, such as *Spergula subulata*, are met with.

3. South Gothland includes all the country lying to the south of the mountains of Smoland. Its regions are more maritime than any others, being open both on the east and west to the sea ; they are almost everywhere girdled by pine woods, except toward the northern boundary, where the beech, with the honeysuckle and elder, take their place. The limits between eastern and western plants are very indefinite here, but the western character generally prevails. In common with West Gothland it contains *Gnaphalium arenarium, Anemone palustris, Inula Pulicaria, Stachys arvensis, Sison inundatum, Holcus mollis, Digitaria sanguinalis*, &c.

The western part is remarkable for the abundance of Leguminous shrubs, such as Genistas, the woods being chiefly beech covered with herbaceous lichens, while *Galium saxatile* is found in elevated rocky spots and *Ranunculus hederaceus* in muddy places near the sea. The most southern angle called Scania is very flat and fertile, and yields *Trifolium striatum, Crepis biennis*, &c., at the S.E. corner ; while on the eastern shore occur loose sandhills, and within them exists a flat and warm region, more protected from the winds and enjoying more of the sun's heat than any other province, so that several southern plants, such as *Alsine viscosa, Euphorbia Cyparissias* and *Asarum europæum* are common.

The eastern islands exhibit a characteristic vegetation : for Gothland, of calcareous formation, produces the Orchises in perfection, and approaches in character to the limestone Alps of Austria : thus *Helianthemum Fumana, Inula ensifolia, Epipactis rubra,* occur there ; and in its bogs, *Orchis palustris, Schœnus nigricans,* &c.; while its extremity possesses, in common with the extreme promontory of Scania, *Cyperus fuscus, Sium Falcaria, Liparis Loeselii, Linaria Elatine,* &c. &c. *Coronopus didyma* and *Ranunculus hederaceus* come here from the western ocean.

Œland is a narrow island, the southern extremity of which descends as a kind of stony desert towards the sea, and is characterized by the presence of *Helianthemum œlandicum, Artemisia laciniata, Thlaspi perfoliata, Ranunculus illyricus, Chrysocoma Linosyris.* To these are added in more fertile spots, *Ulmus effusa, Viola persicifolia,* and various *Verbasca.* The Baltic species, *Chenopodium hirsutum* and *Selinum lineare,* are met with there ; and the Scanian *Valeriana dioica, Veronica triphyllos, Holosteum, Corydalis cava,* &c.

With regard to the proportions of the families of the flora, the following statistics may be given. Sweden north of the lakes contains about 1256 species, 944 Dicotyledons, and 312 Monocotyledons. Among these the largest orders are—

Compositæ $\frac{1}{12}$, Boraginaceæ $\frac{1}{74}$, Labiatæ $\frac{1}{37}$, Scrophulariaceæ $\frac{1}{26}$, Umbelliferæ $\frac{1}{38}$, Ranunculaceæ $\frac{1}{30}$, Cruciferæ $\frac{1}{22\frac{1}{2}}$, Caryophyllaceæ $\frac{1}{25}$, Rosaceæ $\frac{1}{20\frac{1}{2}}$, Leguminosæ $\frac{1}{27}$, Ericaceæ $\frac{1}{63}$ (a comparatively small proportion to Lapland), Salicaceæ $\frac{1}{33}$, Orchidaceæ $\frac{1}{44}$, Juncaceæ $\frac{1}{48}$ (again comparatively few), Cyperaceæ $\frac{1}{13}$, Graminaceæ $\frac{1}{13}$.

Of the families occurring in South Sweden or Goth-land, with its islands, are wanting, Verbenaceæ, Globu-lariaceæ, Santalaceæ, and Ulmaceæ.

Gothland or the S. Swedish peninsula has the richest flora; its species amount to 1416, exceeding those of Denmark; the Monocotyledons are 347, and the Dico-tyledons 1069.

The proportions of the chief orders are :—

Compositæ $\frac{1}{11\frac{1}{2}}$, Boraginaceæ $\frac{1}{70}$, Labiatæ $\frac{1}{28}$, Scrophu-lariaceæ $\frac{1}{25}$, Umbelliferæ $\frac{1}{27}$, Ranunculaceæ $\frac{1}{29}$, Cruciferæ $\frac{1}{22\frac{1}{2}}$, Caryophyllaceæ $\frac{1}{23\frac{1}{2}}$, Rosaceæ $\frac{1}{16}$, Leguminosæ $\frac{1}{19\frac{1}{2}}$, Eri-caceæ $\frac{1}{83}$, Salicaceæ $\frac{1}{44}$, Orchidaceæ $\frac{1}{40}$, Juncaceæ $\frac{1}{64}$, Cy-peraceæ $\frac{1}{15\frac{1}{2}}$, Graminaceæ $\frac{1}{12}$.

Thus, compared with Norway and Sweden, the pro-portions of the Labiatæ, Scrophulariaceæ, Umbelliferæ, Rosaceæ, Leguminosæ, and Orchidaceæ are much in-creased, while compared with those regions, and more especially with Lapland, the Ericaceæ, Salicaceæ, Jun-caceæ and Cyperaceæ are much diminished.

The orders Verbenaceæ, Globulariaceæ, Santalaceæ occur in this region alone; while Tamaricaceæ, Eleagneæ and Lobeliaceæ occurring further north are not found here. The neighbouring country of Denmark wants Glo-bulariaceæ, Polemoniaceæ and Fumariaceæ, while it pos-sesses the order Ambrosiaceæ in excess, *Xanthium* which was formerly found in Gothland being now extinct.

With regard to Finland, we find, as might be expected from the remarks respecting its physical conformation and its continuity with Russian Lapland, a richer flora than in Lapland, the excess being in a great measure de-pendent on the presence of more southern forms, since the oak limit is higher on the east than on the west coast of the Gulf of Bothnia. The slight elevation of the mountains excludes many of the alpine species, and the western forms, most abundant in Western Lapland and Norway, are more rare. We have already mentioned, under the head of Russian Lapland, a number of eastern species which do not occur more to the west ; of these, *Ligularia sibirica, Lonicera cærulea, Myosotis sparsiflora, Cœnolophium Fischeri, Actæa spicata-rubra, Rosa carelica* and *Andromeda calyculata* alone seem to extend into Finland. On the whole, the character of the flora appears to approach most to that of North Sweden, with certain species passing up in an oblique line, by the islands of the Gulf of Bothnia, from Gothland in S. Sweden. *Agrimonia pilosa, Ophrys recurva, Leerzia oryzoides, Ulmus effusa* (found also in the islands of the Baltic, where alone *Ul. campestris* occurs), seem to be the only plants found exclusively in Finland. The total number of flowering plants is greater than in Lapland, but less than in either of the other regions, viz. $966 =$ 262 Monocotyledons and 704 Dicotyledons.

The proportions of the principal orders are :—

Compositæ $\frac{1}{11\frac{1}{2}}$, Boraginaceæ $\frac{1}{51}$, Labiatæ $\frac{1}{32}$, Scrophulariaceæ $\frac{1}{20\frac{1}{2}}$, Umbelliferæ $\frac{1}{36}$, Ranunculaceæ $\frac{1}{26}$, Cruciferæ $\frac{1}{18}$, Caryophyllaceæ $\frac{1}{21}$, Rosaceæ $\frac{1}{20\frac{1}{2}}$, Leguminosæ $\frac{1}{25}$, Ericaceæ $\frac{1}{48}$, Salicaceæ $\frac{1}{37}$, Orchidaceæ $\frac{1}{37}$, Juncaceæ $\frac{1}{56}$, Cyperaceæ $\frac{1}{11}$, Graminaceæ $\frac{1}{12\frac{1}{2}}$; thus in the Orchidaceæ and Ericaceæ it is relatively rich like Lapland ; it has the highest proportion of Cruciferæ, but in most respects it approaches much to Norway and N. Sweden.

The islands adjacent to the Scandinavian peninsula on the east, south and south-east present some singularities which have been already alluded to, but a few more particulars may be added.

The island of Bornholm, situated between the extremity of the Scandinavian peninsula and the continent, affords several species not found on the former, although on the whole it is clearly related to Gothland, as *Mentha rotundifolia, Medicago ornithopodioides, Spiranthes autumnalis* ; several others are common to Bornholm and Gothland exclusively, as *Petasites frigida, Imperatoria Ostruthium, Oxalis corniculata* and *Medicago minima.*

The islands of Œland and Gothland, already alluded to as exhibiting peculiarities, afford the following species not elsewhere found : *Artemisia rupestris* and *laciniata, Linosyris vulgaris, Galium rotundifolium, Verbascum thapsiforme,* and some of the varieties of *phlomoides, Globularia vulgaris, Plantago minor, Ranunculus illyricus* and *nemorosus, Thalictrum Kochii, Adonis vernalis, Braya supina, Coronopus didyma, Helianthemum Œlandicum* and *Fumana, Viola pratensis, Potentilla fruticosa, Kochia hirsuta, Anacamptis pyramidalis, Orchis laxiflora, Iris pseud-Acorus* var. *citrina, Tofieldia calyculata, Carex tomentosa, nemorosa* and *obtusata,* and *Calamagrostis montana.*

The relations of S. Sweden to the continental flora are also shown by a comparison between Gothland and Denmark, which latter does not possess so many species by about 100 ; the proportions of the families, however, are nearly identical.

The following species occur in Denmark, and either in Gothland and the Baltic islands alone, or as far as the south of Finland, of the Scandinavian provinces : *Pyrethrum maritimum, Cineraria palustris, Senecio sarracenicus, paludosus* and *erucæfolius, Gnaphalium arenarium* and *luteo-album, Filago germanica, Inula Helenium* and *bri-tannica, Pulicaria vulgaris, Petasites alba* and *spuria, Echinops sphærocephalus, Cirsium acaule, Sonchus palustris, Tragopogon porrifolius, Picris hieracioides* (to the S. of Finland), *Arnoseris minima, Xanthum strumarium* (extinct in Gothland), *Sambucus Ebulus, Valerianella dentata, Galium tricorne, Campanula Rapunculus, Mentha sylvestris, Prunella grandiflora, Stachys Betonica* (to the S. of Finland), *Teucrium Scordium, Ajuga reptans* (to Finland), *Verbena officinalis, Vinca minor, Verbascum thapsiforme* and *phlomoides* with their varieties (var. *collinum*, Schrad. of the latter alone occurring in Norway and N. Sweden), *Scrophularia Ehrharti, Antirrhinum majus, Linaria cymbalaria, Veronica montana* and *triphyllos, Orobanche elatior* (to the S. of Finland) and *O. minor, Primula elatior* (Jacq.), *Anthriscus vulgaris, Peucedanum Oreoselinum* (to Finland), *Falcaria Rivini, Bupleurum tenuissimum, Acer campestre, Reseda luteola, Corydalis cava, Fumaria capreolata, Arabis arenosa* (to Finland), *Nasturtium officinale* (to Finland), *Lepidium latifolium, Malva moschata* and *Alcea, Geranium palustre* (to Finland), *Oxalis stricta* and *corniculata, Hypericum humifusum, Dianthus proliffer, Holosteum umbellatum, Sedum purpureum, Vicia dumetorum, Ornithopus perpusillus, Tetragonolobus maritimus, Lotus uliginosus* and *tenuifolius, Trifolium alpestre, striatum* and *filiforme, Ononis campestris, Sarothamnus communis, Genista pilosa, tinctoria, anglica* and *germanica, Ulex europæus, Carpinus Betulus, Orchis militaris* (to Finland), *laxiflora, pyramidalis, Leucojum vernum, Allium carinatum, Anthericum Liliago* and *ramosum, Juncus obtusiflorus, atratus, maritimus* and *pygmæus, Arum maculatum, Cyperus fuscus, Cladium Mariscus,* &c. &c.

All these plants are distinctly of southern origin, and are characteristic of the flora of central Europe, many even of the warmer parts.

Thus, regarded as a whole, the flora of the Scandinavian peninsula is made up of plants having the most diversified relations, in each province manifesting features of resemblance toward the countries nearest to which they lie, and which features become less and less marked towards the centre of the region, where floras of the different provinces, each bringing its contributions, meet and become in some degree blended together. The borders washed by the sea in the greater part of their extent exhibit the plants of the countries to which the different lines of coast face ; thus the Icelandic *Gentiana involucrata* and *serrata*, *Kœnigia islandica*, *Primula finmarkica*, *Saxifraga Cotyledon*, and the var. *grœnlandica* of *Saxifraga cæspitosa*, strongly characterize the northern coast above and around the islands of Loffoden, while to the north-east these give place to the Russian *Phaca frigida*, *Senecio nemorosus* ; more to the south to numerous representatives of the British flora, such as *Ligustrum vulgare*, *Pyrethrum Parthenium*, *Sedum anglicum*, and quite to the south *Ulex europæus* and *Glaucium luteum*. The Baltic coasts afford the German *Dianthus arenarius*, *Astragalus arenarius*, *Anthericum Liliago*, &c.

Moreover, in proceeding from north to south the total number of species continually increases, and transverse lines limiting certain plants, such as the oak, the beech, and so on, occur ; thus the vegetation of the peninsula affords a miniature representation of the entire continent of Europe, in which may be traced almost all the various effects produced by external causes, and it thus offers a most interesting illustration of the complex condition resulting from the peopling of districts from very varied sources, under the regulating influences of great diversity of conditions of climate and elevation.

The vegetation may be divided into six classes: 1. The mass of the common species existing throughout Central Europe, and extending to Britain, as well as the Scandinavian peninsula, some of them, however, attaining their northern limit in various parts of Sweden, Norway and Lapland; these belong to the 'Germanic type.' 2. Littoral or maritime plants, generally diffused over the coasts of Europe, but differing in the degree to which they extend northward; some, therefore, reach much higher than others on the Norwegian coast, and others have not yet migrated wholly up the Baltic and Bothnian coasts. 3. The Arctic plants, found exclusively in the northern regions of Europe, Asia and America; these appear to be of varied origin, thus: (a.) many are common to the more eastern portions of the polar regions of Europe and to Siberia, and not occurring in Iceland or Scotland, may be supposed to result from a migration from the east; (b.) others occurring principally on the west side of the Scandinavian mountains, also in Greenland, Spitzbergen, Iceland, or in part in Scotland or its islands. If we adopt the hypothesis of M. Martins in regard to them, we should imagine them to be derived from N. America; but this is yet an open question, and it is quite as probable that their original centre was in western Scandinavia; (c.) thirdly, a certain number of species appearing in Lapland alone (nineteen are mentioned by M. Andersson); there is reason to suppose that the centres of these and of some occurring in the two former categories exist in Lapland itself, to which a portion are as yet restricted, but from whence some of the

foregoing may have migrated either to the east or west.
4. A large proportion of the Scandinavian plants are
' Alpine-boreal,' occurring again in the Swiss Alps : the
positions of the centres of these are involved in great ob-
scurity ; they will be alluded to hereafter, when speaking
of the origin of the British vegetation. 5. There remain
a considerable number of species occurring in the most
southern part of the peninsula or its islands, mostly also
in the neighbouring country of Denmark, and belonging
to a type more southern in its character than the ' Germanic ' (or rather more ' south-western,' since the lines
of vegetation depending upon heat run from N.W. to
S.E. in Europe) ; these are such plants as *Orchis militaris* and *ustulata, Helianthemum Fumana, Verbascum
phlomoides, Geranium rotundifolium, Medicago minima,
Trifolium striatum*, and others, which, occurring both
in the north and south of France, and only abundant in
the southern parts of Germany, might be termed the
' French type.' 6. Lastly, there exist in the S.W. parts
a small number which are found in greatest abundance in
the western parts of Europe, but are not of particularly
southern character ; they occur frequently in the west
regions of Europe and in the north-west of Germany, but
are especially prominent in Britain, whence we may distinguish them from the southern type we have called
' French ' by the name of the ' British type ;' examples
of these are *Ulex europæus* and *Erica Tetralix*.

Sect. 2.

Iceland.

This large outlying insular appendage of the continent of Europe is situated between $63\frac{1}{2}°$ and $66\frac{1}{2}°$ N.L., and extends from north to south about forty-five geographical miles, and sixty-five from west to east, the north-western portion forming a peninsula connected with the rest of the island by a narrow isthmus. The greater portion of Iceland is mountainous, and the elevation is greatest at the south-east, on which side the mountains fall off abruptly to the sea. In this region lie the highest points, Oræfa-Jokull 6400 feet, and Austrojokull 5673 feet high; Hecla, lying in the south-west, attaining 5347 feet. These mountains are thus less elevated than those of Scandinavia. The land sloping most gradually towards the north and west, the rivers principally flow in those directions, and on those coasts occur numerous bays or fiords, which are less frequent on the east and south.

The mountains of Iceland are mostly volcanic, many of them being still active, throwing out lava, ashes and stone, and also water. Sand and loam are sometimes cast out in such quantity as to cover large tracts of land. The warm springs, such as the Geyser, are well-known; they are natural springs, heated by volcanic action to nearly or quite the boiling-point, and emitting the water in fountains rising 100 or 200 feet, probably through the pressure of steam collected in subterranean cavities.

The temperature of the island is comparatively mild for its latitude; the winters are especially so in the southern part.

	Lat.	Ann.	Winter.	Summer.
Reykiavik	64°	41°	$29\frac{1}{2}°$	55°
Eyafiordur	66	32	21	$45\frac{1}{2}$

The mean temperature of Reykiavik is thus about the same as that of the same latitude in Norway, and 6° higher than the mean heat of Umea in Finland. The winter is $6\frac{1}{2}°$ milder than in Drontheim, and nearly 16° more so than in Umea ; but it is proportionately longer, for the mean temperature is below the freezing-point for five months, viz. from November to March. On the other hand, the summer temperature is more than 7° lower than in Drontheim, and not even so high as that of Umea. A comparison of Reykiavik with Eyafiordur leads to the assumption of a great difference between the north and south parts of Iceland. The mean temperature in Eyafiordur is no higher than that of the North Cape, lying 5° more to the north. The great difference of temperature probably depends partly on the mountain range separating the two parts of the island, and partly on the drift-ice which is driven by north-east currents against the east coast of Greenland, and from thence on to the north coast of Iceland. It often remains here until June or July, finally floating off, on the east side, into the Atlantic Ocean. Compared with similar latitudes in North America the temperature of Iceland is still very high, as is seen by a glance at the isothermal map.

The atmosphere is on the whole damp and foggy ; the weather changeable.

In Scandinavia the snow-line sinks down towards the sea, because mists and clouds interfere with the melting. Iceland being still more exposed to the maritime influences, the snow-line is found still lower ; it is met with at 2500–3000 feet. Glaciers extend down from the enormous snow-masses of the mountains into the valleys and even into the sea.

The vegetation of Iceland agrees very closely with that

of Norway, but the trees are almost wanting, only the birch and the mountain-ash (*Pyrus Aucuparia*) occurring, and these attaining but very small stature. On the mountains grow the dwarf birch, and the same low willows that flourish on the Norwegian chains, and the Iceland and Reindeer mosses, with other lichens, are very abundant.

The absence of trees is to be attributed, not to the temperature, evidently, but to the damp, foggy sea-air, the violent storms and variable weather; since great woods occur in Siberia and North America in places where the annual and summer mean temperatures are far more unfavourable. In Scandinavia even, the limit of the zone of forests sinks towards the sea, and the outer islands and most exposed mountains are completely devoid of woods, and the same effect is evident even on the west coast of Jutland.

The cultivation of grain, which is uncertain in the Feroës, is impossible in Iceland. The only plant grown on a large scale is the potato, and of this the tubercles scarcely arrive at more than the size of a large nut. In gardens the inhabitants grow cabbages, turnips, leeks, &c., and beetroot, which does not become much larger than a radish. Barley is not excluded from Iceland by the severity of the winters, for it ripens at Elvbaken, in the north of Scandinavia, under a latitude of 70°, where the mean winter temperature is 20° Fahr. And the summer temperature of Iceland is pretty nearly equal to that of the Feroës and Shetlands, and higher than at Elvbaken. It is clear therefore that the ripening of grain is prevented by the peculiar constitution of the atmosphere in spring and autumn, the spring retarding the growth of the halm, the autumn not allowing the development of starch in the seeds, which always remain gorged with watery juice. The spring of Elvbaken is indeed still colder than that of Iceland, but the summer rains are not so frequent; for while Reykiavik has fifty-one days of rain from May to September, only twenty-one occur at Elvbaken in the same interval. Thus the barley rots as it stands in Iceland; at Elvbaken it ripens incompletely, requiring to be dried by artificial heat. In addition to this, the mean of the *maximum* temperatures is highest at Elvbaken, the soil is relatively drier, and moreover at 70° of latitude the sun is longer above the horizon in summer than at 64°, which must have an important influence upon the development of the plants. It should be remarked, however, that Elvbaken is a kind of oasis in the Lapland desert, for the regular cultivation of barley on the coast of Norway does not extend above 66° N.L.

The only conifer that grows in Iceland is the dwarf juniper, which scarcely attains, in its creeping growth, a length of more than 18 or 20 inches, and is almost exclusively confined to the clefts of the soil, or the irregularities of the streams of lava. The shrubby willows are very common, from the bottom of the valleys to high up upon the mountains. The goat willow (*Salix Capræa*), which grows spontaneously on the banks of the rivers, often gives the aspect of a field of lucerne to the tracts it covers. Other species growing near the snow-limit are creeping and black, with the leaves so small as to be almost imperceptible at a distance, so that they look like masses of roots rather than true shrubs. Some kinds, however, become quite giants compared with their congeners, especially when growing in sheltered situations, attaining a height of 6 feet and more.

The white birch (*B. alba*) is commonly dwarfed and creeping; but in the interior of the island it emerges from this condition, and grows usually in small clumps (seldom singly, or, if so, prostrate), the height amounting to about 6 feet. The Icelanders honour these collections of shrubs with the name of forests. It appears, however, that larger trees must have existed formerly in Iceland, since trunks 15 or 20 feet long are found partially buried in the debris of the volcanos.

When the conditions are too severe for the white birch, the dwarf birch (*B. nana*), which usually grows intermixed with it, wholly takes its place, forming the sole covering of the soil. The heather, whortleberry, and bear-berry clothe the rocks wherever the birches are wanting. Sometimes these grow associated with the dwarf birch, forming a complete carpet around the basaltic rocks of the coast; and a dwarf willow often joins these, or replaces one of them, on the open heaths so common in Iceland, and so often inundated by rains.

According to M. Martins, the supposed disappearance of trees in Iceland is at least much exaggerated. It has been attributed to the incursion of ice from Greenland; but since the vegetation of the north coast is quite as rich, if not richer, than that of the south, it cannot have resulted from the severity of cold produced by the coming of this drift-ice to the coast, and, if true, is probably more attributable to the recklessness of the inhabitants in destroying it without care for the future.

Ferns are extremely abundant in Iceland, since the peat-bogs, where they usually flourish, occupy a large proportion both of the bottoms of the valleys and the slopes of hills which are always moist.

The marsh marigold (*Caltha palustris*) will give a fair

K

idea of the general humidity of the soil, when we say
that it flourishes most prosperously both in the open
marshes and on the roofs of the huts which are con-
structed of earth. It seems to delight in the neighbourhood of the houses, and its golden flowers pleasantly
break the monotony of the places it inhabits.

The best fodder grows, in fact, on the thick walls of
these very farm-houses, or on the tumuli and walls of the
cemeteries, and those of the temple in the centre ; thus
it is not uncommon to see the horses and cows browsing
upon the houses. The only tolerable pasturage is within
the limits of these homesteads, which may be regarded
as so many oases in the midst of the vast bogs which
often stretch from the foot of the mountains to the sea-
shore. In these spots the inhabitants cut all the hay
they can obtain, in the beginning of August, with a short
hand-scythe, of about two fingers' breadth, which they
sharpen from time to time upon a piece of basaltic rock
which they moisten with the tongue ; and the crop they
obtain does not equal the aftermath that is mown at
this same period in France.

The herbaceous plants occasionally acquire dimensions
out of all proportion to those of the trees ; that is to say,
the former equal the development they would acquire in
temperate Europe. This is particularly remarkable on
the north side of the island ; and more on the islands
at the bottom of the bays and gulfs than in the actual
centre of the island, and this is probably dependent on
the sea having a higher, more uniform and constant
temperature than the soil of the island in general.

Mostly, however, neither the herbaceous nor woody
plants acquire even moderate dimensions ; many of the
former are not more than an inch high, and the botanist

who seeks them needs a sharp eye and a careful hand to gather them. It is only in the bottom of extinct craters, on their internal walls, or amid currents of lava, and more especially in the clefts and hollows, that a few plants acquire in these natural hot-beds a development equalling that they exhibit in our fields. These are mostly ferns, such as the male fern and lady-fern, the herb-Paris, or *Geranium pratense*. And M.Martins found *A. uva-ursi* in flower in the bottom of the crater of Stadahraun, at a time when the buds were scarcely open at the foot of the mountain, notwithstanding that the former scarcely received a ray of sunlight.

Near the hot springs, so common and abundant in Iceland, a perpetual verdure reigns, contrasting strangely with the almost always sterile and desolate aspect of the country at large ; but it is remarkable that at the flowering period these plants are not more forward than those of the plains and the surrounding hills. The silica contained abundantly in some of these waters seems to exert a detrimental effect ; this is especially noticed at the Geysers, where the silex exists in greatest quantity and purity. The sandy tracts formed by the silex precipitated from the hot waters are clothed but sparingly, chiefly with *Sedum, Parnassia, Filago, Plantago, Epilobium, Prunella, Galeopsis, Stachys* and *Euphrasia*.

Among the other vegetable formations may be noticed the social growths of the wild pansy and dog-violet, embellishing the rocks of the port of Stikkisholmur ; *Epilobium* with *Cakile maritima*, both in great abundance and at a considerable distance from the sea in the plains of black scoriæ ; *Pinguicula*, with its transparent amethyst stalk and flower, accompanies the elegant and

K 2

delicate *Trientalis* in every spot where *Dryas octopetala* displays its fair white flowers.

The Iceland moss (*Cetraria islandica*), which grows in abundance, is chiefly found on a stony soil, almost always damp, on the elevated plains of the centre of the island, and not on the rocks as might have been supposed. The *Cetraria nivalis* is no less plentiful, but grows under very different conditions; it prefers dry places, and literally whitens the black surface of the old lava-streams, so that at first sight they appear covered with snow; just as in Lapland the Reindeer moss, of a yellowish white colour, clothes the soil so completely that it has the appearance of a surface of flour of sulphur.

The flora of Iceland, composed of some 460 species, presents points of remarkable interest in reference to the relations of the floras of the Old and New Worlds, which have been very fully examined in an important essay published by M. C. Martins, on the vegetation of the Feroë Archipelago compared with that of Iceland and of Shetland, and it is in explaining the relations of these islands that the floras will be most conveniently characterized; for the present it will suffice to say, that of the plants of Iceland a considerable proportion consists of species common all over Europe north of the Alps and Pyrenees, which everywhere manifest an indifference to climatic influence within a rather wide range; the rest are either ' Alpine-boreal,' that is, common to the Alps and to the north of Europe, or ' Arctic,' that is, growing only in the most northern parts of Europe or America.

The total number of species of flowering plants, according to the lists of Messrs. Vahl and Babington, is about 414, of which 282 are Dicotyledons, and 132

Monocotyledons, among which the following families present themselves in greatest proportion: Cyperaceæ nearly $\frac{1}{9}$, Grasses $\frac{1}{9\frac{1}{2}}$, Compositæ $\frac{1}{17}$, Caryophyllaceæ $\frac{1}{18}$, Cruciferæ $\frac{1}{19}$, Amentaceæ $\frac{1}{20}$ (Salices $\frac{1}{24}$), Ericaceæ and Saxifrageæ each $\frac{1}{26}$, Rosaceæ about $\frac{1}{28}$, Scrophularineæ $\frac{1}{32}$, Orchidaceæ about $\frac{1}{35}$, Ranunculaceæ, *Potamogeton*, and Polygonaceæ each $\frac{1}{38\frac{1}{2}}$, Leguminosæ, Stellatæ, Onagraceæ and Gentianaceæ each $\frac{1}{46}$, Labiatæ and Umbelliferæ each $\frac{1}{59}$, Crassulaceæ $\frac{1}{69}$, Halorageæ and Primulaceæ $\frac{1}{83}$.— Of 50 natural families occuring in this flora, 16 are represented by only one species. The low temperature and damp climate are strongly indicated by this summary, and still more is this the case with the list of species. They are almost wholly either alpine or bog plants.

The Feroë Islands.

This group is situated between the $61\frac{1}{2}°$ and $62\frac{1}{2}°$ N.L., to the south-east of Iceland, and about equidistant from thence and from Scandinavia, and is politically a dependence of the latter, but physically belongs rather to Scotland, with which the Shetlands form a connecting link. The largest of the many islands are Suderoë, Stromoë, Œsteroë and Vaagææ. They are all rocky; very steep cliffs rising from the sea to a height of 1000 or 2000 feet; the interiors of the islands rise by terraces and terminate in high peaks, of which Slattaretend in Œsteroë, 2881 feet, and Skiellingfield in Stromoë, 2493 feet, are the loftiest.

The climate possesses the insular character in a high degree ; the winter is very mild and the summer damp and cold, the atmosphere exceedingly foggy and the weather very changeable.

M. Martins has given us an account of the cultivated and indigenous vegetation of these islands, which affords a striking picture of the unfavourable circumstances under which they are placed. He states that in a visit to the gardens in the neighbourhood of Thorshavn on the 30th of June, he saw the seedling cabbages only 2 or 3 inches high, and these are cut in October. Lettuces were not an inch high ; parsley, chervil, leeks, spinach, celery, carrots and beetroot were scarcely out of the ground. Radishes were some 3 inches long; they are eaten at the end of August or beginning of September. The clergyman cultivated rhubarb (*Rh. palmatum*) in his garden; the leaves were about 18 inches long and the stalks some three-quarters of an inch in diameter. There were a few beds of strawberries; but they rarely ripen.

Shrubs succeed no better than the herbaceous vegetables. A few raspberry bushes which never yielded fruit, and a hedge of gooseberries and red currants 6 or 7 feet high, were all he saw. The growth of the trees may be judged of by the fact that a birch three years old was 16 inches high ; a hawthorn of the same age no higher. The mountain-ash alone appears able to acquire the stature of a tree, when sheltered by walls, which protect it from the sea-winds. Thus, in the Governor's garden there were three, each about 13 feet high, branched from the very ground and with a crown some 16 feet in circumference ; the trunks, from 4 to 12 inches in diameter, proved that they were very old. But everything about them denoted the struggle with which they maintained their footing; every branch that rose above the wall of the garden was stripped of leaves, or dead; one tree alone some 3 feet high bore two bunches of blossom; the others were all barren. A little dog-wood and an elder some 6 feet high had their trunks dead, and were sprouting from the crown of the root.

It seemed at first, says M. Martins, that the soil and climate of the Feroës were fatal to all arborescent vegetation, but the birch (*Betula alba*) formerly grew there, since it is met with buried in the bogs. An inconsiderate sacrifice of these trees by the inhabitants has caused them to disappear, as in Shetland and in Iceland, where they indeed still exist, though very sparingly, in the interior of the island. The causes which oppose the growth of trees in the Feroës are, the violence of the winds, which uproot the trees and break the young shoots, the constant dampness of the atmosphere, which affects the functions of the leaves, and that of the soil, which rots the roots: if a tree is planted on the rock, the thin layer of vegetable mould is insufficient to fix the roots solidly and it is soon overturned by the winds. Added to these are the low temperature and want of summer heat, and the irregularity of the seasons, which is such that the winter is sometimes so mild that the sap rises in the trees in January and February, and then the rigorous cold blustering winds of March and April arrest all vegetation. And besides all these climatic influences, in the open country the innumerable sheep crop down all growth, and in the towns, the cats, which fill the houses, tear off the bark of every tree, so that their trunks are stripped to 3 or 4 feet above the ground.

Only a sixtieth part of the soil of the Feroës is devoted to cultivation. The robust variety of barley (*Hordeum hexastichum*) called Scotch Big, the potato and turnips are the only plants cultivated on a large scale.

The barley-fields form a skirt rising a little above the level of the sea; but it is rare that the grain fully ripens, and all the seed is brought from Denmark. The extreme limit of the barley is 340 feet above the sea on the south side of the mountains and 200 feet on the north side. This difference is greater than might have been expected, since in these northern latitudes the sun turns so much farther round the horizon that the north face of the mountains receives relatively more rays than in more southern latitudes; but this is quite outweighed by the violence of the squalls loaded with rain that almost always come from the N.W. and level the crops turned towards the north, while those on the south are more frequently sheltered from the wind. The most fertile islands are Suderoë, Waagoë and Œsteroë; in Suderoë barley is sown in April and harvested in the middle of September.

The potato may be cultivated in some sheltered spots to a height of 800 feet above the sea. Rhubarb and *Angelica* succeed very fairly even at 1600 feet; no cultivation can be carried beyond this elevation.

The indigenous vegetation is made up almost entirely of plants also occurring in Britain, but the number of species is comparatively small. Thus from the latest explorations there appear to be only about 200 Dicotyledons and 80 Monocotyledons, of which 16 Dicotyledons and 8 Monocotyledons are absent in Shetland and Iceland, while all but three of these are British, all however occurring in central or northern Europe. The most abundant families are the Ranunculaceæ, Cruciferæ,

Caryophylleæ, Rosaceæ, Onagraceæ, Saxifrageæ, Compositæ, Scrophularineæ, Plantagineæ, Chenopodeæ, Polygoneæ, Amentaceæ (only Salices), Orchidaceæ, Junceæ, Cyperaceæ and Graminaceæ.

In the vicinity of Thorshavn, M. Martins found among other plants, *Ranunculus repens, Bellis perennis, Cardamine pratensis, C. impatiens, Polygala vulgaris, Empetrum nigrum, Viola canina, Oxyria reniformis, Geranium pratense*, &c., on the level land; the raised points were bare of vegetation. Thrift (*Armeria maritima*) in beautiful condition crowned the more arid and exposed hillocks with its tufts. *Silene acaulis* flourished in similar localities; in depressions of the surface occurred *Scirpus cæspitosus, Eriophorum polystachyon, E. vaginatum, Pinguicula alpina, Veronica serpyllifolia, Orchis sambucina* and *Carex cæspitosa*. Opposite Thorshavn is an island called Naalsoe; here the first plants which presented themselves on the sandy shore were *Honckenya peploides, Potentilla anserina* and *Cochlearia officinalis*; in damp hollows *Eriophorum vaginatum, Carex cæspitosa* and *Saxifraga stellaris* were met with. Ascending the steep slopes of the mountains *Nardus stricta* became the predominant plant, almost excluding everything but a few plants of *Ranunculus montanus, Plantago alpina* and *Luzula spicata*. The Phanerogamous plants which grew upon the exposed summit were *Salix herbacea, Carex atrata, Polygonum viviparum* and *Thalictrum alpinum*.

The following plants, observed by Messrs. Trevelyan and Forchhammer in an ascent of Mallingsfall in the island of Videroë, a mountain rising upwards of 2330 feet above the sea, will give an idea of the distribution in height. They ascended on the south-east side and found the following scale of vegetation :—-

At about 1000 feet appeared the first plants of *Salix herbacea*.

At 1400 feet commenced *Dryas octopetala, Botry-chium Lunaria, Thalictrum alpinum, Azalea procumbens, Veronica alpina*.

At 1800 feet *Dryas octopetala* was very common, and *Papaver nudicaule* began to show itself with *Salix arctica*.

At 1950 feet *Papaver nudicaule* was very abundant, associated with *Arabis petræa*.

At 2000 feet they saw *Sibbaldia procumbens* and *Chamalædon procumbens*.

Finally, from 2300 to 2330 feet, at the summit they gathered *Salix herbacea, S. arctica, Empetrum nigrum, Rhodiola rosea, Silene acaulis, Cerastium alpinum, Vaccinium Myrtillus, Polygonum viviparum, Oxyria reniformis, Saxifraga oppositifolia, S. hypnoides, Armeria vulgaris, Sibbaldia procumbens, Alchemilla alpina, A. argentea* and *A. vulgaris β. pubescens*.

The number of flowering species is about 273, of which 293 are Dicotyledons and 80 Monocotyledons. Of these the largest orders are :—

Ranunculaceæ about $\frac{1}{29}$, Cruciferæ $\frac{1}{16}$, Caryophyllaceæ $\frac{1}{17}$, Rosaceæ $\frac{1}{22\frac{1}{2}}$, Compositæ $\frac{1}{14\frac{1}{2}}$, Scrophularineæ $\frac{1}{29}$, Polygonaceæ $\frac{1}{26\frac{1}{2}}$, Juncaceæ $\frac{1}{24\frac{1}{2}}$, Cyperaceæ nearly $\frac{1}{13}$, and Graminaceæ more than $\frac{1}{11}$,— while the Saxifrageæ are only $\frac{1}{34}$, the Ericaceæ $\frac{1}{32}$, the Labiatæ $\frac{1}{45\frac{1}{2}}$, the Boragineæ $\frac{1}{68}$, and the Umbelliferæ $\frac{1}{97}$.

The Shetland Islands.

These lie between the Feroës and Scotland in 60° to 61° N.L., and on the whole are very like the former, but less rocky; the rocks also are not of such abrupt forms; the country is flattish or slightly undulated, and the highest point, Rona's Hill in the island of Maiul, rises only 1500 feet above the sea. The climate closely resembles that of the Feroës, but the winter and summer appear, from the somewhat imperfect data, to be rather colder; the differences between summer and winter are about the same, rather more than 15° Fahr.; giving a maritime climate more equable than in any other part of the northern hemisphere.

The oat (bristle-pointed oat, *Avena strigosa*, Schreb.) is the chief grain cultivated in Shetland, and with it also the variety of barley (*Hord. hexastichon*, Bear or Big) grown in the Feroës. Turnips, clover and rye-grass are also grown with success. Strawberries, red and black currants, gooseberries and a few kinds of apple are the fruits cultivated in Shetland, and perhaps the success of these, which are not produced in the Feroës, depends upon more intelligent management. The only wild berry is that of *Empetrum nigrum*, which is sold in the markets.

Mr. Edmonstone has endeavoured to benefit the islands by the naturalization of trees; he has been most successful with the ash, on account of this tree putting forth its leaves late and losing them early. The sycamore has proved the most robust. The mountain pine (*P. sylvestris* var. *montana*), the horse-chestnut and the white poplar have become tolerably acclimated. The mountain ash, which is indigenous, does not appear to improve

in any way under cultivation. The willows which seem
best to resist this climate, so inimical to the growth of
wood, are *Salix Russelliana, fragilis, cinerea, viminalis*
and *vitellina*. No transplanted pine, whether brought
from Norway or Scotland, has survived more than a year;
the oak, beech and birch perished still more quickly.
Among the shrubs, the *Arbutus mucronata, Cotoneaster
rotundifolia (Uva ursi)*, and the ivy, have alone vege-
tated and given hopes of their being preserved.

With regard to the sycamore, horse-chestnut and ash,
this hope amounts almost to certainty, since at Busta, on
the west coast of Mainland, there are a large number of
these trees in the garden of Mr. Gifford, several a century old and remarkably beautiful.

The indigenous flora of the Shetlands resembles that of
the N. of Scotland, two species only, *Geranium phæum*
and *Arenaria norvegica*, being peculiar to the former.
The general characters and relations we shall speak of
in connexion with the floras of the Feroës and Iceland,
in the comparative view of the vegetation of these regions,
to which we now proceed.

An examination of the floras of Iceland, the Feroës,
and the Shetlands, shows that there exist 146 species,
belonging principally to the Grasses, Sedges, Cruciferæ,
Scrophularineæ and Compositæ, common to all these
islands, and forming as it were the basis of their vegetation.
Out of these 110 are species common all over Europe north
of the Alps and Pyrenees. The remaining 36 are divisible into several groups: 1. a certain number of littoral
or maritime plants, such as *Cakile maritima, Cochlearia
anglica, Armeria maritima, Triglochin maritimum*, &c.;
2. plants of the 'alpino-boreal' type occurring both in
the mountains of Scandinavia and Scotland and in the

Alps and the Pyrenees, such as *Thalictrum alpinum,* *Silene acaulis, Rubus saxatilis, Sibbaldia procumbens,* &c. A single species, *Ligusticum scoticum,* existing in Scotland, Sweden and Lapland, has not been discovered in the mountains of France or Switzerland.

These 'alpino-boreal' plants differ among themselves; in the Alps or Pyrenees they occur in different zones of vegetation. The small number of those which descend into the plains do not advance equally far towards the south. Thus *Comarum palustre* grows on the swamps of the plains of the centre of France, but does not occur at the foot of the Pyrenees. *Vaccinium Myrtillus* is found in heathy places in the forests near Paris, but further south it is only met with at a certain elevation above the sea. These two alone have been observed in the plains of France, the rest only occur at a certain altitude; some are found in the Vosges, where the highest points do not exceed 4800 feet, and are therefore sub-alpine; all are met with in the Alps, at various heights, some which are truly alpine, near the limits of perpetual snow, such as *Silene acaulis, Juncus triglumis,* &c. The respective proportions are—11 maritime and 25 alpino-boreal, 9 of which are subalpine, occurring in the Vosges, and the rest alpine.

The islands under consideration contain no plants which do not exist also on the continent of Europe, and, according to the hypothesis of the distribution of species from specific centres, which we have explained in the introductory chapters, it might be imagined these islands have been peopled from this the nearest mainland; but they form a link of connection between the Old and New Worlds, and since the greater part of their plants occur also in the northern parts of America, it is fair to suppose, as M. Martins does, that both continents may have had a share in furnishing species. That author has argued this question with great ability, and we shall endeavour to give a brief statement of his views.

Iceland being as near to the coast of Greenland as to the Feroës, the 146 plants which it possesses in common with those islands and the Shetlands might, *à priori*, be derived either from Europe or America. Now an examination of the floras of Britain and North America shows that of those 146 species, all occur in Europe, and almost all occur equally in America; but the 110 common plants, so widely distributed in Europe, although occurring in N. America, are for the most part wanting in Greenland. Now it must be supposed that the American species of Iceland are derived from Greenland, and therefore it seems most probable that these 110 species are derived from Europe, since only 37 occur in Greenland; the other 73 exist only on the other side of Baffin's Bay, in Labrador, Canada, or in the United States, far more distant than the coast of Scotland from Iceland.

Supposing even that more complete observation should raise the number of Greenland species common to Iceland, &c., to 60, there would still be in these islands 50 very common European plants which do not occur in Greenland, and it would be most natural to suppose that these had passed from Europe across the chain of islands continued to Iceland. With regard to three there can be no doubt, since they have never been observed in any part of N. America, *Galium uliginosum*, *Melica cærulea* and *Scabiosa succisa*; the last a very conspicuous plant, and one concerning which there can be no confusion of nomenclature, for the genus *Scabiosa* has no representative in N. America.

The littoral or maritime plants exist equally on both shores of the Atlantic, and therefore offer no evidence either way ; the same is true of the 25 of the alpine and boreal species, which with one exception, *Galium saxatile*, occur in the arctic regions of the New World. M. Martins, however, inclines to the belief that they are derived from the coast of Greenland, since the mountains of Scotland form a secondary centre of distribution, and their flora resembles that of the arctic regions much more than that of the South European mountains.

It has been observed that of the numerous Saxifrages characterizing the alpino-boreal vegetation, *Saxifraga oppositifolia* is the only species common to Iceland, the Feroës and the Shetlands, since in fact it is the only species occurring (and only in two localities) in the Shetlands. Now there are 7 species in the Feroës, 15 in Iceland, and 17 in Greenland ; moreover 5 of these species occur in Scotland, so that 5 species common to Greenland, Iceland, the Feroës and Scotland are wanting in the Shetlands.

The first four of these, *S. stellaris, nivalis, rivularis* and *cæspitosa*, are essentially of the arctic type. In Spitzbergen they all reach the 80° N.L., and their southern limit in Sweden is at 63° on the mountains of Jemtland ; in Norway at 68°, 3 still occurring on the sea-coast at Loffoden, but *S. nivalis* only at an altitude of 1200 feet*. Since these plants stop therefore at 63° in Scandinavia, it is not surprising that they have their southern limit in the Feroës at 62°. In the mountains

---

* M. Martins is in error here : these four plants were found by Prof. Grisebach as far down as 60° N. L. in Norway, the first in Hauglefjeld, the rest in Haidangerfjeld and Folgefonden. According to Koch's Synopsis, *S. nivalis* occurs also in the Riesengebirge.

of Scotland and Wales they again find a suitable climate, cold summers, and abundant snow in winter to preserve their young shoots from the frosts of spring.

It is more difficult to explain the case of *S. hypnoides,* for this is not an essentially arctic plant; it does not occur in Spitzbergen, nor in Scandinavia, nor even in Switzerland, but it is found in the Vosges, descends to the sea-coast on the east side of Scotland, and occurs here and there in Britain as far as the 51°. It is therefore a sporadic species, depending on local conditions for its maintenance, escaping from the general laws of distribution.

If then it be admitted that the Feroës form the southern limit of the arctic species, just mentioned, it will not be surprising that *S. oppositifolia* advances beyond them into the Shetlands, since, while they are arrested at 68° on the west coast of Norway, the opposite-leaved Saxifrage occurs there at the sea-level in 67°. In Switzerland it comes down to the limit of the silver-fir (*Abies pectinata*), that is to say, about 5000 feet; on the west of Scotland it grows on the coast, and advances along the lower mountains of Wales as far as 53°. It is one of those plastic species which accommodate themselves to great climatic modifications and very different stations. In Switzerland it occurs on arid and denuded rocks, from which the wind often strips the snow in the depth of winter. It was the first plant observed by M. Martins in Spitzbergen, growing almost at the level of the spring tides; it requires the smallest amount of heat of almost any known plant, for M. Bravais observed it in flower at Alten in Lapland on the 5th of May, 1839, while *S. cæspitosa* did not flower till the 11th; and at Hammerfest, further north, *S. stellaris* and *rivularis* were not in flower till June 29th, and *S. nivalis* till July 2nd of the same year. The absence of those four species is therefore a result, not of the higher temperature of the summer, but the want of protection in winter and more particularly in spring ; while the more robust *S. oppositifolia* resists the icy winds and spring frosts which would kill its congeners.

The Feroës lie much nearer to the Shetlands than to Iceland ; the climate of the two archipelagoes is almost the same, while that of Iceland is more severe. On the other hand, the Feroës are full of mountains, several of which exceed 2000 feet, and exhibit three zones of climate, diminishing respectively nearly 2° in mean temperature as they rise. These climates therefore approximate to that of Iceland, like the physical character of the country and the geological structure of the soil.

The statistics of the vegetation demonstrate that the Feroës belong rather to Iceland, while the Shetlands are an appendage of the Scotch flora ; namely, 67 plants are common to the Feroës and Iceland, and are unknown in the Shetlands ; only 37, common to the Feroës and the Shetlands, are wanting in Iceland. These last are of the common continental species of Europe ; all occur in Britain, only 5 are wanting in the environs of Paris. They are thus of the same physiological constitution as the 146 species common to the Shetlands, the Feroës and Iceland, but are not quite so plastic, and so are unable to pass the limit of the Feroës. One plant alone of the species common to the Feroës and the Shetlands does not exist in the plains of Europe, *Cerastium latifolium*, L. This plant is common in the high Alps ; it occurs in the mountains of Scotland above 3000 feet. It does not grow at Spitzbergen or in Lapland, and has only been

L

found in one locality in Norway, in the Dovrefjeld, but it has been discovered in Greenland. It belongs therefore to the alpino-boreal type; but the predominance of the European type in the Shetlands, and the occurrence of this plant in the intermediate group of the Orkneys, lead M. Martins to regard it as derived rather from Scotland than from Greenland.

The great majority therefore of the plants common to the Shetlands and the Feroës have been derived from Europe. Several, indeed, belong to *genera* unknown in N. America, *e.g.* the daisy, *Erica cinerea, Anagallis tenella, Littorella lacustris* and *Scilla verna.* Of species there are 18 which do not exist in America, such as *Stellaria graminea, Hypericum pulchrum, Iris pseud-acorus, Holcus mollis,* &c.; the remainder occur in both continents, but the greater part not north of the United States, and are kinds which would be expected to have become naturalized in Iceland if they had passed across it, such as *Lychnis dioica, Achillæa ptarmica, Taraxacum palustre, Eriophorum vaginatum, Dactylis glomerata,* &c.

It has been said that 67 species exist in Iceland and the Feroës, but not in the Shetlands. This is almost double the number common to the two archipelagoes, and yet Iceland is far more distant than the Shetlands, from the Feroës. We have to seek an explanation of this apparent anomaly. Among these there is a certain number which belong to an eminently arctic type; such as *Ranunculus nivalis, Draba verna, Papaver nudicaule, Saxifraga nivalis, S. rivularis, S. tricuspidata, S. hypnoides, Cornus suecica, Angelica archangelica, Kœnigia islandica, Salix lanata* and *S. Lapponum.* If we study the distribution of these plants in western Europe, we arrive at the following results: *Papaver nu-*

dicaule exists only at Spitzbergen*; *Ranunculus nivalis* is confined to northern Lapland†; *Saxifraga nivalis* and *rivularis*, so common in Spitzbergen and around Hammerfest, 70°·40 N. L., do not advance south beyond the 68°‡; *Saxifraga tricuspidata*, Retz., is a Scandinavian plant§; the rest are also north-European species. All are found equally in Greenland, and have come from thence to Iceland, thence to the Feroës, where several have found their southern limit, not reappearing even in the mountains of Scotland, as for instance, *Ranunculus nivalis, Papaver nudicaule, Saxifraga tricuspidata* and *Kœnigia islandica.*

After these come naturally those plants which grow both in Scandinavia and the Alps, the alpino-boreal, 28 in number, almost all inhabiting the region bounded by the eternal snow; among them 18 occur between 8933 and 8943 feet above the sea on the terminal cone of Faulhorn, in the canton of Berne. The others do not occur in this locality, but are essentially alpine.

It is therefore not surprising that the 34 species of these two series should stop at the Feroës, and only reappear on the mountains of Scotland, where several find a suitable climate, and in particular the shelter of deep snow in winter.

The origin of these plants is to be sought in N America; in fact, out of the 28, only 2 are absent from the list in ' Flora Boreali-Americana,' namely *Hieracium Lawsoni* and *Ranunculus glacialis.* The first is a critical species, and there may be some error in the determination; but the second is a distinct and essentially alpine species, growing on lofty mountains near the snow-

* Found also by Grisebach on Dovrefjeld.　　† Ibid.
‡ 60°. See *ante.*　　　　　　　§ Not given by Fries.

line. Thus it is very rare in Spitzbergen ; it appears in
the northern mountains of Scandinavia : the Sylfjeld, a
mountain of 6600 feet in the 63rd parallel, is its most
southern limit in that country. In Greenland it has only
been detected on the shores of Baffin's Bay. It is not
unlikely that it will be found on the summits of the
mountains of Iceland or the Feroës.

The remainder of the plants unknown in the Shetlands,
but common to Iceland and the Feroës, belong neither to
the arctic nor the alpino-boreal type. Of these, 12 are
aquatic, such as *Myriophyllum verticillatum, Montia
fontana, Potamogetons,* &c. These are not very sensi-
tive to climatic influence ; they all exist both in Europe
and America ; but their presence is intimately connected
with the depth, purity and freshness of the waters which
they inhabit. Since these conditions occur in the Shetlands, it is not evident why these species should not have
established themselves there, if they had passed, by the
Shetlands, from Britain to the Feroës ; it seems more
likely therefore that they have invaded the Feroës from
Iceland ; but why they have not passed on to the Feroës
is inexplicable.

Among the remaining 15 species are several which
have probably been introduced by man, with the grain
brought from Denmark for the support of the inhabitants
of Iceland, and for the annual sowing in the Feroës ;
these are 5 in number. Then there are two ferns, *Asplenium Trichomanes* and *Cystopteris fragilis,* plants flourishing in the crevices of rocks, but not on cliffs exposed
to the maritime blasts. It is not surprising therefore
that they do not occur in the Shetlands, for they only
find their suitable stations on the mountains of Iceland
and the Feroës.

Lastly, there are 8 plants of which it is impossible to explain the absence in the Shetlands, occurring, as they do, in Scotland, the Feroës, Iceland and N. America.

A certain number of plants occur in the Shetlands and in Iceland which have not been met with in the Feroës. These, amounting to 40, are all continental or maritime species widely distributed in central Europe ; 21 are found in the environs of Paris. It is difficult, however, to conceive how they can have passed from the Shetlands to Iceland without having become naturalized in the Feroës ; for they must have found on the sea-coast there a climate analogous to that of the Shetlands, and in the mountains one analogous to that of Iceland : now, as they are capable of multiplying in both of these stations, it is not evident why they are deficient in the Feroës. But of these 40 species, 34 grow in the most northern part of America, to which Iceland is so near ; it might be supposed, then, that they have come directly from America to Iceland. But at the same time it is improbable that they have crossed the Feroës, and therefore M. Martins concludes that the Shetlands have derived them from Europe ; and in fact a trace of such an European migration is preserved in 5 plants unknown in America, which have settled in the Shetlands and Iceland without fixing themselves in the Feroës. These 5 are, *Sinapis arvensis, Hedera Helix, Erica tetralix, Juncus lamprocarpus,* and *Potamogeton lanceolatum.* There might be some confusion in the determinations of the two last, but there can be no doubt attaching to such plants as the field mustard, the ivy, and the cross-leaved heath ; especially as N. America has no species of *Erica* or *Sinapis.* These three plants do not exist in America nor in the northern part of Scandinavia, therefore M. Martins thinks that they must have passed gradually from Britain and the Shetlands to Iceland without becoming acclimatized in the Feroës.

The derivation of *Trientalis europæa*, an extremely common Scandinavian plant, is considered doubtful by M. Martins ; he believes that it has reached Iceland and the Shetlands from Norway, as also *Arenaria norvegica*, but their absence from the Feroës cannot be explained.

In the Shetlands occur 90 species which are found in Europe, but not in the Feroës or Iceland. Of these 74 are common European plants, but more than half (40) are wanting in America, among which are the white water-lily, the hawthorn, and many others about which there can be no mistake. This large proportion goes to prove that the continental plants occurring only in the Shetlands have been derived from Europe, and, more sensible to the severity of the winters and the absence of heat, have attained their northern limit there. Of 13 maritime plants, 9 are wanting in America, and have therefore been derived from Europe ; the other 4, common to the old and new worlds, have probably had the same origin. Of the three boreal species which close the list, two have never been found in America, but all are common in Scandinavia, where they have been derived, either by direct migration or by way of Britain ; *Geranium phæum* pointing to a direct migration, since it does not occur truly wild in Britain. Another species, *Saussurea alpina*, occurs on the Scotch mountains, but is not common below 67° N.L. in Norway. It has not been found in Greenland, but only on the Rocky Mountains and at Behring's Straits on the west side of America. It has therefore been derived from Scandinavia. The third species, *Cherleria sedoides*, is doubtfully given

as a native of the Shetlands, and it does not occur on the
Scandinavian peninsula or in Greenland. Although not
rare on the Scotch mountains, it does not appear in the
Orkneys ; if it be really a native of the Shetlands, it is
an alpine species, derived from Scotland, and attaining
its northern limit in the Shetlands.

In the Feroës have been found 31 species which as
yet have not been met with in Iceland and the Shetlands. Of these 26 occur both in Scandinavia and in
Britain. Two other Scandinavian species appear to be
absent from Britain, although they exist in France ; one
of these, *Atriplex hastata*, is a critical species ; the other,
Orchis sambucina, not so. *Alchemilla argentea*, Don,
and *Draba rupestris* occur in Britain, and *Alchemilla
fissa* will probably be found there. Thus, of 31 species
occurring in the Feroës and not in the Shetlands or
Iceland, only 4 are at present wanting in British lists.
All are continental plants occurring in Central and Northern Europe. The majority also exist in America ; but
12 have not yet been ascertained to exist there. This
seems to show that these plants have come to the Feroës
from Europe rather than from America ; and M. Martins
considers that all belong to the European migration,
from the British Isles to the Feroës through the Shetlands, and that the reason they have not become naturalized in the last is to be found in their hostile climate,
where the winds of spring are loaded with icy particles
and freeze the young shoots, which there are not protected by the mountains and the long-enduring snows
they find in the Feroës.

In Iceland occur 135 plants which are not found in
the Feroës or in the Shetlands. One portion of them
grow in Central Europe, 72 in number ; another portion

amounting to 30 species are alpino-boreal; and the remaining 33 are arctic.

Of the 72 plants growing in central and northern Europe, 53 are common also to America, 19 wanting. This seems strange at first, as we should expect the exclusively European plants occurring in Iceland and not in the Feroës or Shetland to be very few; as would be the case if Iceland had derived its plants wholly from America, and in particular from Greenland. Therefore the existence of the exclusively European species proves the European migration to have extended as far as Iceland, and it is probable that they have been derived directly from the coast of Norway, since it is inconceivable how 21 French species could pass through the Shetlands and Feroës without becoming naturalized in their archipelagoes. This view is confirmed by the study of the second section. These 30 alpino-boreal plants include 5 species unknown in America, but very common in Scandinavia. It is true they exist also in the Swiss Alps, but it is more natural to suppose that they have been derived from Norway than from central Europe, especially when one of them, *Arenaria ciliata*, only occurs in Ireland, and *Orchis nigra* not at all in Britain; while if they had passed to Iceland by the Scotch mountains, we should expect to find them naturalized there. The last section of the list comprehends 33 exclusively arctic species, common to the polar regions of both continents. All occur in Greenland whence they have been derived, for the absence of any exclusively European species forbids the supposition of their derivation from Scandinavia.

Thus, to sum up the views of M. Martins in a few words, he considers that the two archipelagoes and Iceland have derived their plants from two migrations, an

European and an American. The former, much the most considerable, has brought the common European species, of which, although many occur in N. America, not one is peculiar to it; it has also furnished the alpino-boreal; and the migration appears to have been of a double kind, principally by way of Britain, but partly by a direct passage from Norway.

The American migration has probably furnished the exclusively arctic species, some of which do not occur on the European continent, and of those that do, many occur in Iceland which are wanting in the Feroës and Shetlands, through which they would naturally pass to Iceland from Norway, where moreover they grow at a great elevation and are less common than in Greenland which is so much nearer to Iceland.

The European migration seems to be clearly proved, but sundry objections occur to the hypothesis of the American migration. There does not appear to be good evidence of the existence of above three or four plants in Iceland and the Feroës which are not found elsewhere in Europe, and these may yet be found in Norway. The coast of Greenland is icebound, and the only known region in the Danish settlements appears unlikely to have furnished an American contingent, on account of its inland position in Davis's Straits, and the fact that its endemic characteristic plants are not common to Iceland. The nearest American coast from which an immigration might have been produced by the Gulf-stream is Labrador, but this is far more distant than Norway from Iceland, and the Icelandic arctic plants are not Labradoric but Norwegian in character, corresponding with those of the Fjelde about Bergen, i.e. the nearest point in Europe.

Therefore there seem to be equally good grounds for

believing that the arctic plants as well as the rest may have migrated from Europe, and their probable centre would then be Norway, which would thus be considered a primary centre of distribution. These plants may have migrated directly to Iceland, and from thence to America.

The speculations which have just been detailed lead to many highly interesting questions, but it is exceedingly difficult to come to any very definite conclusions at present. The field is but newly opened, and we require a much larger body of data than we possess at present ; the results which have been arrived at hitherto must be regarded rather as the fruits of experimental trials of the ground than as steps solidly secured upon it, but they afford evidence of the existence of sufficient material for widely extended and most curious investigations.

Sect. 3.

The British Islands.

GREAT BRITAIN and Ireland lie between the 50° and $58\frac{1}{2}°$ N.L., and both have a greater diameter from north to south than from east to west, so that their forms are elongated, though this is much less the case in the latter. The length of Great Britain amounts to 120 geographical miles, of Ireland only to 60 ; the breadth in the former varies from 15 to 65, and that of Ireland from 20 to 40. The general outline of the west coast of Ireland, although indented by bays, is rounded, while that of Great Britain is deeply excavated, the bays in the north, in Scotland, being especially narrow and deep. Numerous islands lie on the west side of Scotland, and a few, especially the large Isle of Man, between England and Ireland ; but there are none of any importance on the west side of Ireland or the east of Great Britain.

Scotland, or North Britain, is divided into three portions by two natural boundaries. The first of these is formed by two deep bays, the Murray Frith and Loch Linnhe, connected by a chain of lakes, little elevated above the sea, Loch Lochy, Loch Ness, &c., and by the Caledonian Canal, constituting as it were a narrow fissure running from N.E. to S.W., and separating the north of Scotland from the central portion. The second boundary running more E. and W. is formed principally by the Friths of Clyde and Forth and a tract of low land lying between them.

The mountain masses of Scotland stretch from N.E. to S.W., two in the central region and one in each of the others. The north-eastern part of the north division is flat, but the remainder highland, the highest points of which lie toward the south-west. The most elevated point is Ben Wyvis, 3708 feet. In central Scotland the northern range is formed by the Inverness mountains, the highest point being Ben Nevis, attaining 4385 feet, the greatest altitude in the British Islands. The second range is the Grampians, the highest point of which, Ben-Muich-dhu, reaches 4320 feet, Cairn-Gorn 4067 feet. These two groups are higher and more precipitous on the west side, and are separated from each other by Lochs Awe, Lydoch and Ericht, and the river Spay, forming a line also running from N.E. to S.W. The mountains of South Scotland are lower, rising towards the N.E. ; the highest point is Hartfell, 3289 feet ; they have a similar direction to the others.

In England the mountains are on the west side and

present three distinct masses : one in the northern counties, in which Helvellyn rises to 3215 feet and Skiddaw to 3012 feet; a second in North Wales, with Snowdon 3587 feet ; and a third, much lower, in the S.W., where the highest point of Dartmoor reaches 1786 feet. The eastern part of England is generally level ; ranges of low hills vary the middle and southern portions.

The Irish mountains are mostly isolated ; they occur especially in the vicinity of the sea ; the interior of the country presenting merely low hills or level plains. A tract of level land extends across Ireland from Dublin to Galway Bay, nowhere rising above 300 feet. The mountains of Killarney are the highest summits in Ireland : Gurran Tual is 3404 feet, Mangerton 2754 feet; in the Slievebloom ridge, some exceed 2000 feet; among the hills of Mourne, Sleithdonard is 2796 feet, and Errigal in Donegal 2462 feet.

Thus, the mountains of Scotland are higher than those of England ; those of Ireland are the lowest, while all are inferior to those of Scandinavia. As in that country, they are chiefly collected and steepest towards the west. The mountains of Scotland are mostly composed of primitive rocks, limestone occurring only in the south ; they agree therefore in the main geological features with the Norwegian. In England the primitive rocks only protrude in the west ; all the east is secondary and tertiary ; limestone is abundant, and in the E. and S.E. chalk.

In Ireland both primitive and more recent rocks come to the surface. Some of the Hebrides, part of the west coast of Scotland and of the north coast of Ireland, consist of basalt, as in the island of Staffa and the Giant's Causeway.

The temperature may be concluded from the following figures :—

	Lat.	Ann.	Winter.	Summer.
Edinburgh ...	56°	48°	39°	59°
Dublin.........	53½	49	40	59
London	51½	50	38½	62
Penzance	50	51	42	61½

A comparison of the temperature of Edinburgh with that of Copenhagen, in nearly the same latitude, shows the annual mean of the former to be more than 1°, and that of the winter nearly 8° higher, while the summer heat is more than 5° lower. Thus Scotland, as might be expected, has a much more insular climate than Denmark, and the same holds good of the British Islands generally. In Dublin the difference of the seasons is much less pronounced than in London, which is less exposed to the influence of the ocean, and the distinction is still smaller in Penzance, where the temperature of April equals that of Copenhagen, while the summer heat does not rise so high as that of Stockholm.

The climate is moist and rainy, more especially in the west and north, and the atmosphere is not so clear as on the continent. From a mean derived from observations made in a number of places, the annual rain amounts to 28–30 inches ; but in some places it rises to 60 and over. The greatest fall of rain occurs on the western coast and on the west side of the mountains or elevated chains of hills in general. The number of rainy days (including snow-falls) is considerable ; in Dublin 208, in London 178, while Copenhagen has only 134. The greatest quantity of rain falls in summer and autumn,

but the winter is also rainy ; much less snow falls here
than in equal latitudes on the continent, and it seldom
lies in the S.W. of England or in Ireland.   The mountains nowhere reach the snow-line, but the highest summits, such as Ben Nevis, Snowdon, &c., are covered with
snow during the greater part of the year, and isolated
patches of it are met with here and there in the summer.

Through the comparatively mild climate resulting
from their maritime position, the British Islands possess
a more southern vegetation than countries situated in the
eastern parts of Europe in the same latitude, while the
inequalities of its surface affording a great variety of conditions of temperature, allow of variety of forms surpassing that of most of the northern countries of Europe.
The distribution of the plants of England and Scotland
has been very thoroughly investigated ; that of the Irish
flora is not so well known, yet the principal features
have been so far determined that the general character
is well understood.

The great body of the indigenous vegetation is made
up of plants common to the North German, or Central
European flora : these extend over and characterize the
lowlands throughout, excepting in the south-west confines of England, and more especially of Ireland, where
the much higher temperature, the moisture, and possibly
certain other circumstances, give rise to the presence of
forms spreading down the western coast of Europe as far
as Spain.   The mountains again, and in the northern
provinces even the comparatively low tracts in their
vicinity, are distinguished by an approach to the peculiar
arctic character, and the presence of many species common
to Scandinavia and the higher mountains of southern
Europe.

In a highly-populated country like Great Britain it is obvious that the long continuance of cultivation, and the other accompaniments of civilized life, must have in a great measure destroyed or altered the more striking landmarks of an indigenous vegetation, in particular the forests and the plants peculiar to the richer and less elevated districts. Thus the few existing woods of great extent are scarcely to be regarded as natural products, since they are mostly subject to the repairing care of man; but still, planted woods, indicating the quality of climate by their flourishing condition, are equally available with the indigenous forests as evidence of existing physical conditions.

The forests of Scotland are chiefly composed of the Scotch fir or pine and the birch, but the oak is also abundant, and in the south the beech occurs; Edinburgh may be regarded as the northern limit of this tree in Britain (56°), not so high as that in Scandinavia, but still a more northern one than in the east of Europe. In England and Ireland the oak and the beech are the prominent trees of woods; but as the limit of the latter includes those of many of the other catkin-bearing trees, and also the ash, the maple, &c., and the plantation of trees having been long a favourite pursuit, our woods, so frequently artificial, exhibit a very great variety. Perhaps we may say that the oak is the most characteristic forest-tree in general; while the country within a circuit of fifty miles round London, and some of the western counties, exhibit a more considerable number of beech-woods. The chestnut also thrives in woods of the south of England (about 51°), and the hazel is universally distributed in Britain.

The northern parts of Scotland and the mountainous

parts of the north of England present large tracts of moorland, open heaths, or bogs mostly devoid of trees; while the lowlands are throughout distinguished by the peculiar luxuriance and freshness of the turf, resulting from the moisture of the climate. Peat-bogs are especially abundant in Ireland, somewhat less so in Scotland and the north of England, but are gradually diminishing in various parts from the advance of cultivation.

The general character of the climate favours the growth of all plants which do not require a very high temperature to ripen their fruits and seeds; hence the successful cultivation of green and root-crops, rendering Britain so essentially a grazing country. The hardier fruits are common, but the more delicate only succeed in the southern parts; thus neither the mulberry nor the walnut ripen in Scotland, while the peach and apricot require care and protection in the south, and the grape can hardly be said to be an out-of-doors plant in England. Its place is taken by the hop, chiefly cultivated in the south-east; and the apples, so largely grown for cider-making in the western and south-western counties of England. Yet although the mulberry, walnut, peach and grape fail in the north, while they ripen freely in the open air in Denmark, the bay-laurel and many other introduced southern plants stand our winters commonly in the southern counties, and in Devonshire the myrtle, the fuchsia, the camellia, which must be wintered in conservatories even in central Europe, are exposed to the climate throughout the year with impunity.

Wheat is the prevailing bread-corn, barley being cultivated for malting, and oats for horses; rye is not nearly so common. In Scotland wheat and barley prevail in the east, and oats in the Highlands and in the west.

The potato predominates over grain in Ireland ; oats are next in abundance, wheat and barley following. But in the highly elaborate agriculture of the British isles numerous other crops find place ; besides the grain and root crops of the rotations, the clover, turnip, mangold, &c., beans and peas, are commonly grown ; in Ireland flax ; here and there in Britain buck-wheat ; while the operations of market-gardening are carried to the highest perfection. In regard to the indigenous vegetation, we have a tolerably perfect knowledge of it, due in great measure to the labours of Mr. H. C. Watson, who has devoted so much time and energy to the study of the distribution of British plants. In his 'Cybele Britannica' we find the latest statement of his conclusions on this subject, and of these we shall avail ourselves largely in the following brief sketch. The indigenous or naturalized plants of Great Britain amount to between 1400 and 1500 (no definite number can be given, since opinions as to specific value and established naturalization differ), consisting of species in almost all cases common to some other part of Europe or North America, and variously distributed within these islands. Any attempt to classify these geographically can be only approximative, since in most cases the examples of each species are so widely distributed that particular tracts present only less or more of them. They may however be considered in three ways : 1. in reference to their horizontal distribution over the country ; 2. in reference to the altitudes at which they especially occur ; and 3. in reference to their affinities to the plants of neighbouring countries.

Mr. Watson divides Great Britain into eighteen botanical provinces, the boundaries of which are founded upon physical and not political differences ; the basins of rivers,

M

tracts otherwise presenting a naturally definite character, forming the provinces, which are thereby rendered more natural areas than the accidentally divided counties. It would lead us much too far into detail to attempt to indicate the characters of these provinces, and this is the less necessary as the general characteristics of the different parts of Britain are exhibited under the other two heads above mentioned.

The zones of vegetation depending on altitude are characterized by Mr. Watson differently from the usual mode, since the trees are not safe guides in a country like Britain, where their absence must necessarily be the result of accidental or rather artificial causes. The cultivation of a crowded county is carried up the sides of the mountains as far as the climate will allow, and careful observations upon this afford a line of demarcation between the two principal regions, the arctic and the agrarian. The highest point at which Mr. Watson has observed the cultivation of grain in the Highland mountains is at the outlet of Loch Callater, estimated at 1600 feet above the sea. Potatoes can scarcely be grown in Drumochter Pass, calculated to be 1530 feet above the sea, and much more shadowed by the mountains. From 1000 to 1200 feet is the more common limit of corn and potatoes in the Highlands. The common brake fern (*Pteris aquilina*), distributed throughout Britain, is found to be limited by a line running nearly level with the limit of cultivation, and thus affords a test, when cultivation may be absent where nature does not deny it success. In one sheltered spot in the woods of Loch-na-gar it was observed at 1900 feet, and in another part of the same woods at 1700 feet; but on the exposed moors it is very seldom seen beyond 1200 feet, unless in hollows or on declivities facing the sun.

These two regions are again divisible into zones, characterized by plants limited to one of the great divisions, but unequally diffused within this. Making allowance for the local variations always occurring where physical characters and geographical position are different, the following plants are taken as characteristic of the zones of altitude in Britain, as determined by Mr. Watson.

1. The Super-arctic, bounded below by the limit of the heather (*Calluna vulgaris*) at an elevation of about 3000 feet; above this line occur comparatively few flowering plants; at the summit of Ben-muich-dhu, one of the highest of the Grampians, exceeding 4000 feet, occur *Silene acaulis, Carex rigida, Festuca (vivipara?), Luzula arcuata* and *spicata, Salix herbacea* and *Gnaphalium supinum*; but besides these alpine plants others are met with which come up from the very lowest region, such as the marsh marigold (*Caltha palustris*), the bog violet (*V. palustris*), the thrift (*Armeria maritima*), and the sweet vernal grass (*Anthoxanthum odoratum*), together with numbers from the next zone; so that this region is defined rather *negatively*, by the absence rather than the presence of particular plants; *Saxifraga cernua* and *rivularis*, however, are perhaps peculiar to it.

2. The Mid-arctic zone is that in which the botanist meets with the richest treasures of rare alpine plants. It lies between the limit of the heather (*Calluna vulgaris*) at about 3000 feet, and that of the cross-leaved heath (*Erica Tetralix*) at about 2000 feet. Here grow those little plants distinguished by their bright-coloured and conspicuous flowers, the *Gentiana nivalis, Veronica alpina,* &c., with *Saxifraga nivalis, Erigeron alpinus, Astragalus alpinus, Alopecurus alpinus. Trollius europæus* runs into this region, as also the spotted orchis (*O.*

maculata), the common harebell (*Campanula rotundi-folia*), and the milk-wort. Moreover the Scotch fir (*Pinus sylvestris*) occurs in the lower parts of this region and the juniper. Many of the common alpine plants grow in all three regions, and this, as standing intermediately, is thus the richest in the arctic forms. Some, such as *Thalictrum alpinum, Alchemilla alpina, Saxifraga stellaris, Oxyria reniformis, Polygonum viviparum,* &c., descend some distance into the Agrarian region.

3. The Infer-arctic region is limited above by the *Erica Tetralix,* about 2000 feet, below by the brake fern and the limits of cultivation, about 1000–1400 feet. Here, of course, the plants approach more closely to the lowland character, a number of species creeping up from the plains, often suffering a stunting of their habit from the more severe climate; the birch too makes its appearance here, growing probably generally throughout this region.

These three zones of the arctic region are characterized generally by the affinity of their flora to that of the most northern parts of Europe, and in a less degree to that of the higher parts of the Swiss Alps.

> " The rugged mountain's scanty cloak
> Was dwarfish shrubs of birch and oak,
> With shingles bare, and cliffs between,
> And patches bright of bracken green,
> And heather black, that waved so high,
> It held the copse in rivalry."

They occur in greatest development in Scotland, including all the mountain-ranges rising above 2000 feet; they are represented again in the north of England, on the mountains of Cumberland and Westmoreland, on the moors and higher hills of Yorkshire, and in North Wales. In Ireland there occur some additional species common to its west coast and the Pyrenees, among which certain species of Saxifrage may be mentioned. The mountains on that coast also present still more strikingly than those of Scotland, the tendency of alpine plants to descend towards the sea-coast.

4. The Super-agrarian zone, or highest region of cultivation, is stated to comprise three-fourths of the surface of Great Britain, including : 1. all the coast-line and low plains or moors in the north and north-west of Scotland, where plants of the alpine character descend even to the sea-shore ; such as *Thalictrum alpinum, Draba incana, Saxifraga oppositifolia, Arbutus alpina,* and *Dryas octopetala*; 2. all other spaces in Britain where the elevation of the ground leads to the production of the same or usually associated species, such as *Arbutus Uva-ursi, Saxifraga stellaris, Alchemilla alpina, Tofieldia palustris,* and *Juncus triglumis*; 3. those tracts of slight elevation upon which a corresponding flora and general vegetation prevail, apparently in consequence of mere proximity to high mountains ; *Saxifraga aizoides* growing so low as 300 feet among the mountains of Cumberland, and *Epilobium alsinifolium* at 500 or 600 feet in Caernarvonshire, while they are never seen at so low an elevation in England in situations remote from the higher hills. This zone is moreover distinguished from all the arctic zones by the presence of the holly, hazel, oak, ash, honeysuckle, hawthorn, and the bramble ; and from the lower zones by the absence of certain species named below.

5. The Mid-agrarian zone comprehends all low grounds, clear from mountains, situated between the estuaries of the Clyde and Tay on the north, and those of the Humber and Dee on the south ; probably also a narrow coast-line of the eastern Highlands, extending

from Perth to Aberdeen, possibly to Inverness. To these must be added a narrow belt winding round the hills of Wales characterized principally by the vegetation of this zone. It is distinguished from all the zones above it by the presence of the black and white bryonies (*Bryonia dioica* and *Tamus communis*), the maple, the common and alder buckthorns (*Rhamnus catharticus* and *Frangula*), the dwarf furze (*Ulex nanus*), the mealy guelder-rose (*Viburnum Lantana*), the spindle-tree (*Euonymus europæus*), and the dogwood (*Cornus sanguinea*). Unlike the mid-arctic zone, it presents few or no species which can be regarded as entirely restricted to it. It contains species coming down from above, such as *Trollius europæus, Geranium sylvaticum, Habenaria albida, Rubus saxatilis*, and some others, which are very rare, if not altogether absent from the lowest zone.

6. The Infer-agrarian zone embraces all the country southward of the Dee and Humber (continued into the Trent), excepting the higher hills and moors of the botanical provinces of the Severn and the Peninsula. It is especially characterized by the greater abundance of the species commencing in the zone above. Among the most conspicuous and attractive species absolutely restricted to it is the wild clematis or traveller's joy (*Cl. Vitalba*), which is especially luxuriant in the calcareous tracts of the southern, eastern, and inland counties. The wild madder (*Rubia peregrina*) is another characteristic plant most abundant in the southern and western counties. Several species, also quite peculiar to this lowest zone, are too scarce or local to serve as fair characters, such as *Erica ciliaris, Sibthorpia europæa, Cyperus longus,* and *Scilla autumnalis*, which are confined to circumscribed localities in the south and south-western counties.

The characteristics may be summed up as follows, reversing the order of the figures in accordance with the usual mode of tracing the plants upwards.

Arctic Region.

6. Upper arctic zone—the herbaceous willow without the heather.

5. Middle arctic zone—the heather without the heath.

4. Lower arctic zone—the cross-leaved heath without the brake fern.

Agrarian Region.

3. Upper agrarian zone—the brake fern without the buckthorn, &c.

2. Middle agrarian zone—the buckthorn without the clematis.

1. Lower agrarian zone—the clematis, madder, galingale, &c.

Unless the summit of Snowdon be supposed to reach the upper arctic zone, none of Mr. Watson's 18 botanical provinces will include all six zones. In other places the hills are not high enough to surpass the middle arctic zone, until we arrive at the Highland provinces, all three of which have their summits above the limits of the heather, but their lowest portions near their coast-line and southern boundary fall barely within the mid-agrarian zone.

The infer-agrarian zone occurs at the coast-line from the mouth of the Humber all round the east, south and west coasts of England and Wales up to Anglesea. The mid-agrarian runs on the west from Anglesea to Argyleshire, on the east from the Humber to Forfarshire ; and the super-agrarian zone lies at the coast-line all round the north of Scotland and its adjacent islands.

This is necessarily only a meagre outline of the characteristics of British vegetation, since our space is limited, but we have the less to regret since the works of Mr. Watson, and more particularly the concluding volume of the 'Cybele,' shortly to be published, will contain an accurate and extended account of the subject, such as we could not pretend to give here. We therefore refer those interested in pursuing this subject to Mr. Watson's writings.

With regard to the distribution of Irish plants we have less definite information, but one or two local floras having been published for that country. It is most distinguished by the absence of alpine conditions, none of the mountains attaining very great height ; and these being fully exposed to a maritime climate, the vegetation displays little or none of those differences at different elevations which are met with in most cases. Another important point is the presence of a certain number of plants in the S. and S.W. which belong to the Pyrenean flora, and are not met with elsewhere in the British Isles or in Northern Europe. To this point we shall refer again presently.

The British Flora is of very complex character, and admits of division into a number of groups. These are naturally much intermixed in a confined space like our islands, but their relations to the floras of other parts of Europe point to a diversified origin.

1. The mass of British vegetation is made up of plants common all over central Europe, which are conveniently denominated plants of the 'Germanic' type ; but among those which have been included by geographical botanists

under this name, there appear to be plants differing to an important degree in their range of distribution. Besides the common plants, a part of the Germanic type are western, others eastern in their character, being limited on one or other side of the continent of Europe by lines which denote their dependence either on the maritime or the continental climate ; that is to say, some are favoured by moisture and milder winters, others are indifferent to the cold, but require great summer heat. A further portion appear dependent rather upon the length of the day, and therefore the amount of direct sunlight, than upon the mean temperatures, and thus are limited by lines coinciding with the parallels of latitude. Thus we get four sections of Germanic plants, viz. : *a.* the universal Germanic plants, such as *Ranunculus acris, Glechoma hederacea, Lamium album* and *purpureum, Bellis perennis,* &c. : *b.* the mid-Germanic, the general limit of which is marked on the north by the 52nd and 53rd parallel, but which extend sporadically beyond this ; such as *Phyteuma orbiculare, Ophrys apifera, Ajuga Chamæpitys* (sporadic), *Lithospermum purpureo-cæruleum,* &c. : *c.* the west-Germanic, favoured by the oceanic climate and mild winters, such as *Corydalis claviculata, Fumaria capreolata, Hypericum Elodes, Malva moschata, Genista anglica, Erica Tetralix, Ulex europæus, Narthecium ossifragum,* &c. : and *d.* the east-Germanic, of which few examples occur in Britain ;—two probably recently naturalized plants, *Stipa pennata* and *Bupleurum falcatum,* belonging to this section.

The members of the first section, the universally distributed Germanic species, are arrested at various points in advancing north ; some do not reach Scotland, such as *Myosurus minimus, Scleranthus perennis, Anemone Pul-*

satilla, &c. ; others do not appear to extend to Ireland, such as *Diplotaxis tenuifolia, Stellaria nemorum, Valeriana dioica, Scabiosa Columbaria, Campanula glomerata,* &c.

2. There exist a large number of British plants which occur principally in the south-western and southern counties ; these belong to what, in speaking of the Scandinavian flora, we have termed the French type, examples of which are : *Matthiola sinuata, Senebiera didyma, Rubia peregrina, Erica vagans, Sibthorpia europæa, Cicendia filiformis,* &c. ; all these occur also in Ireland, but some belonging to the same type do not extend thither, as *Corrigiola littoralis, Erica ciliaris, Lobelia urens, Scilla autumnalis,* &c. These are all related particularly to the flora of the N.W. of France and the Channel Islands, but are not peculiar to the coast of France, extending as far as the valley of the Rhine about the Moselle district. They are greatly intermixed with what we have called the west-Germanic section, and it would be difficult to draw a very marked line between them ; they seem however to require a moister climate, and belong rather to easily disintegrable soils, yet like the west-Germanic are sometimes found on compact rocks, such as limestone or chalk.

3. In Ireland are found a certain number of plants, occurring on the west and south-west coast, which are not found again until we arrive at the north of Spain ; there are 6 Saxifrages, *S. umbrosa, S. Geum,* &c., *Erica Mackaiana* and *mediterranea, Dabœcia polifolia, Arbutus Unedo, Pinguicula grandiflora* and *Arabis ciliata* ; and perhaps some others. These are called the Iberian type.

4. We have the mountain plants, which in this point

of view may be divided into two sections : *a.* The boreal type, occurring in the north of Europe, but not in the Swiss Alps ; such as *Draba rupestris, Saxifraga rivularis, Arenaria rubella, Astragalus alpinus, Phyllodoce cærulea, Primula scotica,* &c. ; all of these occur on the highest Scotch mountains, none of them on the Welsh. *b.* alpino-boreal, occurring both in Scandinavia and the Swiss Alps or Pyrenees ; some of these are truly alpine, such as *Veronica alpina* and *saxatilis, Lychnis alpina, Betula nana, Sedum villosum,* &c., which are found on the Scotch mountains and in part on the Welsh. A large number of subalpine plants, common to Norway, Scotland, and the Southern Alps, as well as to the Vosges and other mid-European subalpine chains, belong here ; these range downward from the higher subalpine regions to the south of Scotland and north of England, and include such plants as *Corallorhiza innata, Ajuga pyramidalis, Linnæa borealis, Cornus suecica, Trientalis europæa,* &c. An isolated case occurs in *Saxifraga nivalis,* which is found in Lapland and Norway, in Iceland and the Feroës, in the Scotch, the English and the Welsh mountains, and only in the Riesengebirge on the continent of Europe ; so that it is more limited on the south than the true alpino-boreal plants.

5. Lastly, there are a few sporadic plants occurring in one or two localities, of which two, *Eriocaulon septangulare,* found in the Western Islands of Scotland, and *Sisirynchium anceps,* found in Ireland, belong exclusively to the N. American flora ; and *Naias flexilis,* said to occur near Dantzig also, is another Irish plant which may have been derived from N. America. Many other cases might be cited of sporadic occurrence of evidently exotic species, but those are mostly to be explained by

direct human agency, as by transport in ballast, &c.; but the case of *Eriocaulon* seems to point to the influence of the Gulf-stream, and this may have brought across some of the other peculiar and local species.

SECT. 4.

The North-European Plain.

UNDER this name is comprehended the large flat expanse of country bounded on the north by the Baltic, the North Sea, and the Channel separating England from the Continent, and on the south by several groups of mountains, of which the Hartz and the Weser mountains are the most northern; to the east and west of these two ranges, the elevated country retreats more and more from the coast, and thus the plain becomes wider at each extremity; but there, however, a projection is formed by the Danish peninsula, which, as well as the Danish islands (excepting Bornholm), Rügen, and some small ones lying off the north coasts of Holland and France, must from their natural characters be combined with this region.   On the west the boundary is formed by the Atlantic Ocean, on the east the plain passes immediately into the East-European; a boundary may be fixed by the river Niemen and the region of the sources of the Dnieper and Dneister.

On the west this plain lies between 46° and 49° N.L. (the west coast of France); in the middle, between 52° and 58° (from the Hartz to Skagen); and on the east between 50° and 55° (from the Carpathians to the embouchure of the Niemen). According to political divisions, it includes the north of France, Belgium, Holland, North Germany, Denmark, Prussia, and Poland.

Although this tract is flat, and on the whole uniform, differences of level do occur. Thus in the N.W. parts rise the range of hills called the Montagnes d'Arrée, having a mean elevation of 500 feet (the highest point 1000); then to the east follows a hilly country (on the N.E. of France and Belgium), rising in the south into the low ranges of the Côte-d'Or, Plateau de Langres, and the Ardennes, the average height of which is about 800–1000 feet, the highest summits reaching 1600–1800 feet. From thence the country is very flat, in part below the level of the sea, and protected from its inroads by dykes (Holland, East Friesland, the north of Hanover, with the west coasts of Holstein and Sleswick). Further east comes a hilly tract, consisting of Denmark and the countries bounding the Baltic on the south. The highest points here are in Jutland and the island of Rügen. To the south of the hills, which form a sort of dam against the Baltic, lie the great North-German sand-plains (Hanover and Brandenburgh), the highest points in which reach only about 500 feet; lastly, between the Weichsel and the Niemen is a ridge, running parallel with the Baltic coast, averaging 350 feet, but reaching 600 at Hafenberg; to the south, the country is flat, or only hilly.

This plain is traversed by important rivers, of which the Rhine rises in the Alps, the rest all in the central European mountains. Lakes occur, especially in Holland, Holstein, Zealand, Mecklenburgh, and in the Prussian hills.

Solid rock crops out but seldom over this tract; primitive formations on the Montagnes d'Arrée, limestone in the north of France and in Belgium, gypsum in Lüneburg and Segeberg, chalk in the islands of Möen and Rügen, and in several other places. Over the remainder of this great plain the soil is composed of sand, loam and other earths, frequently mingled with drift. Peat, still constantly in progress of production, is very abundant in this plain, especially in Holland, the N.W. of Germany, and Denmark.

The temperatures are shown in the following table :—

	Lat.	Annual.	Winter.	Summer.
	°	°	°	°
Paris	49	51	$38\frac{1}{2}$	65
Hamburgh	$53\frac{1}{2}$	48	$32\frac{1}{2}$	64
Copenhagen	$55\frac{1}{2}$	47	31	64
Berlin	$52\frac{1}{2}$	47	31	64
Königsburg	$54\frac{1}{2}$	$43\frac{1}{2}$	26	60
Dantsic	54	$45\frac{1}{2}$	31	63
Warsaw	52	48	30	68

This table, examined independently and again compared with that of the British Islands, shows the gradual decrease of mean temperature towards the east, away from the Atlantic, while the contrast between summer and winter becomes greater (compare isothermal map). The winter temperature of Copenhagen, nearly in the same latitude as Edinburgh, is 8° lower than in that city, the summer temperature more than 5° higher. Warsaw, $3\frac{1}{2}$° more south than Copenhagen, has a colder winter than the latter, but the summer heat is greater than in Paris, although this city lies 3° more to the south. The difference between the summer and winter means amounts in Warsaw to about 38°, in Paris to nearly $26\frac{1}{2}$°, and in London to $23\frac{1}{2}°$. Berlin, 3° south of Copenhagen further distant from the sea, has the same mean temperature.

The quantity of rain is less on the Continent than in the British Islands. A probable assumption gives it at 26 inches in Holland, 20 inches in Denmark and N. Germany. Here, as in Britain, the greatest amount falls in summer and autumn; but less in winter than in England.

The west wind is predominant all over this plain, as is also the case in the British Islands; but its prevalence over the east wind diminishes towards the east. The west wind blows more frequently in England than in Denmark, more there than in Russia. The predominance is most marked in summer; in the winter the easterly winds are almost as frequent as the westerly upon the Continent, which is not true of the British Isles. In most parts the east winds are more frequent comparatively in spring, than in the other seasons.

In the north of France and Belgium the forests chiefly consist of oak and beech. The same holds good of Denmark and the German coast of the Baltic. In Denmark and the adjacent region, the beech is decidedly the most abundant, and this tree seems to flourish better here than in any other part of Europe. In the sandy plains of North Germany the conifers prevail, namely the Scotch fir (*P. sylvestris*), spruce fir (*Abies excelsa*), and silver fir (*Ab. pectinata*), while the birch also is widely spread (many of the forests are planted). In the eastern parts both amentaceous and coniferous forests are met with; here, particularly on the borders of the East-European plain, occur remnants of the great primæval forests. In this part the beech does not reach the northern boundary of the plain, but stops at the 58°, while in Norway it is found as far up as the 59°. The vast heaths are particularly remarkable in this plain; the largest forms a zone stretching through Hanover and the central portion of the Danish peninsula.

From the fact of the lines of vegetation dependent on the climatic influences running at a considerable angle to the parallels of latitude, we find this great tract of northern Europe exhibiting considerable differences as we pass from west to east, as we leave the maritime climate and advance towards the extreme continental condition of eastern Europe. Thus the north-western corner of France, fully exposed to the influence of the Atlantic, manifests a close resemblance to the south and south-west of England and Ireland, and possesses a number of plants which do not occur upon the north coast of Germany, or in central Europe, such as *Matthiola sinuata, Lepidium Smithii, Lavatera arborea, Hypericum Androsæmum, Umbilicus pendulinus, Erica ciliaris, Anchusa sempervirens, Sibthorpia europæa, Polypogon monspeliensis, Lagurus ovatus*, &c.

But the great mass of the plants of the north of France are common to the south also, and there exist only a certain number of plants which find their southern limits on the Loire, these being mostly northern species occurring in N. Britain or Scandinavia. *Cineraria palustris* and *Andromeda polifolia* do not extend south of Rouen. The north-west of France presents a highly favourable field for agriculture, and especially to the feeding of stock, on account of the vigour with which root and green crops and natural meadows flourish. Farming is carried to higher perfection here than in the S. of France; apples and the other hardy fruits are very extensively grown; the vine does not thrive sufficiently to make good wine.

It is in the neighbouring country of Belgium, however, that we must look for the perfection of agriculture, the signs of which indeed become more evident on the French border as we approach the Low Countries. This is not owing to the natural advantages of the country, but to the indefatigable industry of the inhabitants.

The soil, for the most part naturally poor, consists generally of alluvial clay-loams near the coast, and various sands and light loams in the interior. The most fertile soil is that of the lowlands reclaimed by vast systems of dikes and embankments, from the sea; this is composed of muddy deposits mixed with marine shells and fine sea-sand. These lands are called 'polders.' The climate is inferior to that of France or S. Germany; and although the changes are not so sudden as in Great Britain and Ireland, the winters are longer and more severe. The mean temperature of summer is a little higher than that of counties in the same parallel in England, and the time of harvest perhaps a week earlier. The indigenous vegetation of Belgium and Holland closely resembles that of England, the extent of sea-coast giving it a still nearer relation than is possessed by those portions of the present region lying more to the east.

The general character of the coast is pretty uniform up to the Danish peninsula, and we may take a sketch of the vegetation of the islands of Sleswick for an example. High sand dunes are commonly piled up upon the sea-shore, as on the west side of the island of Amrum, facing the open sea, while the two horns of the sickle-shaped island turned towards the main-land run out into fertile marshy tracts, the remaining space being occupied

N

by a treeless ridge covered with heath. These barren high grounds, elevated about the alluvial tracts, are distinguished by the inhabitants by the name of *geest* or *gast*, and seem to represent the original coast, running out in promontories and ridges into the 'marsh' which has been deposited around them ; the *geest* is hilly and mostly sandy, occasionally bearing a few trees; the 'marsh' is always flat, composed of a rich alluvial soil, and wholly destitute of trees, except on the artificial mounds on which the houses are sometimes built ; the '*geest*' has its springs, brooks or rivers ; the 'marsh' is intersected by canals and dikes, but has no springs of its own.

The dunes are consolidated by a cultivated vegetation of grasses and sedges, such as *Arundo arenaria, Elymus arenarius, Carex arenaria,* and *Nardus stricta,* the creeping rhizomes of which form a close network in the loose soil and bind it together ; the bottoms of the hollows more sheltered from the sea are covered with *Empetrum nigrum,* and among this occur a few sand-plants, especially *Dianthus Carthusianorum,* which is not found again southwards until we reach the Rhine. On the outside of the external dunes the sea-bottom falls down almost perpendicularly to a depth of 10 or 12 feet, and the outermost and exposed edge of the land is girdled by a zone of *Zostera, Fucus,* &c.

The 'marsh' is clothed with a rich growth of grass, and bordered by luxuriant vegetation of maritime plants, such as the glassworts, *Salsola Kali, Schoberia, Cakile, Statice Limonium,* and *Aster Tripolium. Salicornia herbacea* occupies the outermost parts, succeeded by the varied groups of sea-weeds clothing the alluvial shore, to where the sand appears from beneath the mud. The '*geest*' was originally wholly covered with heather and

Erica Tetralix, almost the only fuel of the inhabitants, but its sandy surface is now almost everywhere cultivated; some of the maritime plants occur also in the damp hollows here, such as *Aster Tripolium*, and the sea-lavender, which, when in flower, colours large tracts with its blue blossoms.

Passing to a more inland district of the region now under examination, we may find in the wide, elevated moors of the Ems, an example of the character of the level tracts of N. Germany. Like the dry, springless hilly surface of the heaths of Lüneburg, the damp peaty soil of the high moors of the Ems district is everywhere covered with *Erica Tetralix* and *Calluna*. The former is indeed more abundant, but this does not depend upon the moist substratum so much as on the coast-climate which here prevails. The tufts of heath grow on little mounds, a few inches high, more distinctly circumscribed than on the dry ' geest ' where the heather is more densely crowded. The black mud in the interspaces is overgrown with cotton-grass (*Er. vaginatum*) and *Scirpus cæspitosus*, so that tufts of sedges alternate with the islets of heath all over these moors, and where the soil is still wetter the bog-moss (*Sphagnum acutifolium*) makes its appearance. These are the principal forms determining the character of the vegetation; but these moors, so long as they remain in their original condition, scarcely possess above twenty plants, such as *Empetrum, Myrica, Narthecium, Orchis elodes, Andromeda, Drosera, Galium hercynicum, Juncus conglomeratus, Carex panicea*, &c., with a few bog-mosses and lichens. The cultivation of buck-wheat, which is carried on to some extent on these moors, enriches the vegetation with a few more species; and when the soil is again

N 2

relinquished, the original condition is restored but very gradually and never quite perfectly. The natural pools of these moors, the so-called lakes, are destitute of water-plants, but these establish themselves in the pits left by the removal of the peat, when water accumulates in them, such as *Potamogeton oblongus*; but a covering of *Sphagnum* soon extends over, and gradually fills up these pits. When the moors are reclaimed to a cultivation of a higher character, with the feeding of cattle, the vegetation is much more importantly modified, the drained peaty soil acquires a covering of grass and herbs, and soon presents a continuous turf of *Anthoxanthum odoratum*; the fields are then tilled, vegetable-gardens and orchards appear, and even trees thrive to a considerable age; the planted trees are followed by the forest-trees of the surrounding country, and by degrees the barren surface of the plain exhibits woods to the eye of the traveller, who in the unreclaimed districts knows them only as indicating the boundary of the moorland.

For a last example of the character of the vegetation of the North-European plain we may select the province of Brunswick, lying between the Elbe and the Weser. It consists of low plains or hills of little elevation, the highest points, in the southern part, not exceeding 1100 feet above the sea. The mean temperature of spring is about 47°, of summer about 65°, of autumn about 50°, slight variations existing in different parts of the district, especially the southern and eastern, from the differences of elevation, the vicinity of higher mountains, the greater quantity of water evaporated, and the greater extent of the tracts covered by dense forests. The condition of the atmosphere as to moisture is exceedingly varied, but never very damp, on account of the constant movements and frequent changes of wind. The west, north-west and south-west winds predominate, bringing much aqueous vapour, but their violence and alternation with easterly winds so far counteract the tendency to produce excessive moisture, that the rainy days scarcely equal the bright and warm days in the course of a year.

Brunswick rejoices in a vigorous, in some parts luxuriant, vegetation. Above all, the limestone, chalk and sandstone hills are distinguished by splendid woods, chiefly of beech; the sandy and argillaceous hills and plains, where human influence comes to their aid, seem greatly to favour the growth of oaks, birches and firs; the higher parts in the N.W., W. and S.W., S. and S.E., are almost wholly clothed with fine forests. Where the country is flatter and falls off more to the N.W. a greater uniformity of the sandy surface is perceived; the deciduous woods diminish towards the N., N.W. and N.E.; but even there large tracts are covered with noble oaks and beeches; in the plains the birch and fir flourish, and the alder in the low grounds. There also occur large level spaces clothed with a scanty growth of grass and dry heaths; but even these are not unfavourable to the growth of firs, birch, and oak.

On the whole there are few parts which can be said to be wholly irreclaimable to cultivation, these being either steep arid slopes of the hills or undrainable bogs. Elsewhere we find universally either plantations, differing according to the soil, or agriculture, in which are especially distinguishable the meadows, in the humous and moist regions, while the fenny moors produce abundance of alders and of coarse hay.

Among the trees not really indigenous which flourish

luxuriantly, are the walnut, horse-chestnut, sycamore, tulip-tree, mulberry, chestnut, &c.; and a number of more southern species, such as the cedar, plane, *Gledit-schia triacanthos, Populus monilifera,* &c., bear the climate without protection, except in very severe winters.

The apple, pear, cherry, and plum grow in perfection, in infinite varieties, yielding abundantly, except when late nocturnal frosts in May injure the blossom, an accident which happens most frequently in the districts nearest to the Hartz mountains. Peaches and apricots require shelter from the N. and N.E. winds. The vine does well in tolerably sheltered places, the earlier sorts almost always ripening.

The principal grains cultivated are rye, wheat, barley, and oats:—rye mostly as winter rye; wheat almost exclusively as winter wheat; barley mostly as a summer crop; oats as summer oats. Buck-wheat is cultivated profitably in the sandy districts. The leguminous seeds and fodder plants, and root crops, potatoes, flax, &c., all have their place in the agricultural operations; as also hops, tobacco, maize (sparingly, mostly on the borders of fields), and all the usual vegetables of temperate climates in the gardens.

The usual course of the seasons is about as follows: Spring commences towards the middle of March; cherries, plums, and pears and apples blossom towards the end of April, or in the beginning of May; by which time the buds of the forest-trees are strongly developed, and henceforward unfold with almost visible rapidity to clothe the woods with brilliant green, especially when a rain-storm occurs in the earlier days of May. After cold winters, after short storms in March and the beginning of April, there usually follows, in the middle of

the last month, a rapid transition to a spring temperature little below that of summer, which however seldom
lasts more than a few days. In the beginning of May
the sowing of summer crops takes place. About the
middle of June come the earliest peas, some cherries and
strawberries. The hay harvest is in the beginning of
July. Towards the end of July the wheat harvest begins
in the drier and tolerably elevated districts, in those with
poorer soil about the 1st to 4th of August, and near the
Hartz not till the 4th or 8th; but it is over in all parts
by the end of this month. The autumn sowing of grain
takes place in the first-mentioned districts between the
24th and 30th of September, in the last from the 1st to
the 12th of October. Towards the end of September
the leaves turn colour, the equinoctial storms strip the
woods; the last field-crops, and a scanty after-math of
hay, close the season with the end of September or middle of October, and vegetation is at rest from November
to February.

The flora of Brunswick contains from 1000 to 1100
species, almost all identical with those of the midland
and southern portions of Britain, but in part differing,
and exhibiting a relation to the more southern floras of
Germany. The richest families are the Compositæ,
forming $\frac{1}{10}$, the Grasses $\frac{1}{12}$, the Sedges $\frac{1}{16}$, the Leguminosæ $\frac{1}{18}$, including several species not indigenous in
England, such as *Tetragonolobus siliquosus, Galega
officinalis, Trifolium agrarium, Coronilla varia*, &c., belonging to the central European flora; Rosaceæ $\frac{1}{22}$,
Caryophylleæ and Labiatæ each $\frac{1}{23}$, Umbelliferæ $\frac{1}{24}$,
including *Peucedanum Cervaria, Laserpitium latifolium*
and *pruthenicum, Seseli annuum*, not British; Cruciferæ
$\frac{1}{25}$, Amentaceæ $\frac{1}{27}$, Ranunculaceæ $\frac{1}{28}$, Scrophulariaceæ

$\frac{1}{37}$, Gentianeæ $\frac{1}{40}$, Orchidaceæ and Naiadaceæ each $\frac{1}{46}$, Ericaceæ $\frac{1}{66}$, while of the Saxifrageæ there are but 4, or $\frac{1}{265}$, Cistaceæ but 1, or $\frac{1}{1060}$, &c.

Before passing to the examination of the next region, it will be interesting to notice some of the relations between it and the present where they adjoin in the northwestern part of Germany.

The plain of N. Germany rises in a very gradual slope from the North Sea to ridges forming the northern border of the highlands of central Germany. Disregarding a few isolated ranges of hills, the level is from 500 to 1000 feet. This region may also be divided into two distinct terraces, one for the most part lying below the level of 300 feet, and consisting of alluvial deposits; while the other, having an average elevation of 500 to 1000 feet, is formed by stratified rocks. The passage from one to the other is sometimes sudden, sometimes effected by an intermediate and comparatively little developed terrace of about 300 to 500 feet average level.

The lower plain extends over East Friesland, Arenberg and Oldenburg to Bremen, at an elevation of 1 to 100 feet. From thence over the south of Oldenburg, Hoya, Diepholz, Calemberg to Minden and Hanover; and on the east from Bremen over the south of Lüneburg to a line from Buxtehude to Walsrode, and from thence in the valley of the Aller to Celle and Gifhorn, at 100 to 200 feet. Next follows the Lüneburg Heath, between the Elbe and Aller, the *geest* of Altmark, the region between Hanover, Hildesheim, Brunswick and Wolfsburg, and the Westphalian basin between the Teutoburger Wald and the slate mountains of the Lower Rhine, at 200 to 300 feet.

The intermediate small terrace of from 300 to 500 feet

occurs in the border region of the stratified rocks, in the Westphalian hills between Osnaburgh, Minden, Rintelen, Limgo and Bielefield, and between Brunswick, Wolfsberg, Helmstedt and Halberstadt.

The upper terrace, from 500 to 1000 feet, comprises the district between the Upper Weser and the Lein and the zone around the Hartz.

The two terraces possess very different vegetations, but it would be difficult to demonstrate the influence of the climate in this respect, and to distinguish it from that of the substratum; yet in many cases it cannot but be noticed, where plants of the northern part of the plain are confined to high points upon the upper terrace, which is the case with *Trollius europæus, Linnæa borealis, Ajuga pyramidalis, Listera cordata,* &c.; while on the other hand many mountain plants of the upper terrace do not occur on the lower terrace as above defined, and are not found again in the north till we reach Denmark, as is the case with *Aconitum neomontanum, Geranium lucidum, Habenaria viridis,* or Pomerania, as with *Polemonium cæruleum, Salix bicolor,* &c.

The upper terrace is everywhere traversed by rocky ridges, which, according as limestones or argillaceous sandstones predominate, determine the character of the vegetation. The level land north of the hills, the *geest,* the drift on which indicates that it was once a sea-bottom, is covered with a loose soil, mostly sandy, often to a great depth, and thus where it is flat and no watercourses pass through it, extensive peat-bogs are formed. These contrasts are the most important causes of the difference of the vegetation on the two terraces, and would render this still more marked were it not that the sporadic stations of many plants produce abundant local exceptions. But

the local interferences are slight compared with two general conditions contributing to diminish the difference
of the soil of the two terraces, one of which was well known,
the other, however, little observed, until pointed out by
Prof. Grisebach. The *geest* is not merely bounded by
calcareous marshes on the coast and on the banks of the
rivers, but the border next the upper terrace differs essentially from the heaths and moors in the composition of
its soil.

The marshes, increasing products of the present time,
which owe the lime they contain partly to the sea and
partly to the mud washed down by the streams from the
distant mountains, in spite of their favourable soil, have
scarcely a trace of the vegetation of the upper terrace ; but
it is very different on the narrow strip of fertile land which
extends from Osnaburgh over Hanover and Brunswick
to Neuhaldsleben, and through the plain of Magdeburgh
to Barby, and thence far over central Germany, and
which, forming an intermediate step between the upper
and lower terraces, may from its position on the border
of the plain be compared with the marshes. A level surface and rarity of abrupt rocks ally this tract to the old
sea-bottom of the *geest*, like which it is also covered with
drift ; but the everywhere loamy, calcareous soil here produces a most distinct vegetation, compelling us to refer
this tract botanically to the stratified terrace, whence, in
the course of so many centuries, the characteristic plants
have immigrated and established themselves here. Comparison with the marshes is therefore inessential in considering the distribution of the indigenous plants ; but
from their origin we see that this tract is to be considered
as the *marsh of the diluvial* period, formed before the
ocean had retreated from the old stratified coast and left

the existing plains dry. For, just as in the present day the level bottom of the North Sea receives the sandy portion of the detritus carried down by its rivers, while the argillaceous deposit of the marshes is precipitated as on the edge of a filter on the German coasts, so also the *geest* or bottom of the diluvial sea possesses the sandy layers, and the shore-line along the edge of the hilly ridges has received the clay with remains of calcareous shells; the only difference being that human art now wrests from the waves by means of dikes the fertile shores, which in former ages became joined to the mainland and opened to terrestrial organic activity only by a vast natural disturbance. Moreover, as the present marshes are produced in unequal breadth along the shores, and on the duny coast of Holland are buried in sands by stormy movements of the waves, so the diluvial marsh reaches far in the east, from the Hartz to the Elbe, and diminishes gradually towards the west. Near Hanover it is more than nine miles broad, and forms the fruitful tract of Calemberg between the Deister and the Lein ; but in the Westphalian basin, a sea-bottom, it is wholly lost where the heath of Senner approaches closely to the Teutoburger Wald ; a proof that, like the Friesian " polders," it is no immediate product of the rivers, but has been a gift of the partial and inconstant ocean.

The characteristic marks of the diluvial marsh lie in its deciduous woods. Wherever the argillaceous soil occurred, as the sea left the old coast, the Hercynian oak and beech forests advanced downwards from the heights into the plains, while the sandy *geest* became covered with conifers and heath shrubs. Thus we may regard the diluvial marsh now as the cleared site of an old deciduous forest, where agriculture has partially expelled the first inhabitants, which occur now only sporadically, but the more abundantly in proportion as the tract increases in breadth towards the east.

To this region belong the Schaumburg forest near Bückeberg, the Eilenriede near Hanover, the woods around Brunswick, also parts of the Kölbitz and Letzling forests, the vast forests of the plain west of the Elbe, and the grand oak woods on this river between Magdeburg and Dessau, as far as Lausitz. With the trees, also the shrubs and herbs of the heights, which live beneath their shade, descended into the plain, and thus all the woods are more or less rich in the plants of the hills, from which they are almost everywhere separated by cultivated country in the present day. These herbs prove, at the present time, the former connexion. Taking into consideration the slowness and difficulty of immigrations of new plants, which presupposes a dislodgement of a previously existing vegetation, their great variety in these woods is an historical testimony of the antiquity of their residence there.

The upper terrace, on the rocky subsoil, is divided into an eastern and western region by a north-western line of vegetation, common to many plants, extending on from Neuhaldsleben, Halberstadt, Nordhausen, and Eisenach to the Rhine. The east half belongs almost wholly to the basin of the Elbe, the west to the watershed of the Weser. The Elbe terrace has 100 more species than the Weser region, the latter having a much poorer flora, scarcely possessing 20 species not found in the former. This striking contrast does not depend at all upon differences of soil, but almost wholly on the fact that the plants which require a higher summer heat than is received in the region of the Weser, are much more numerous in the north-west of Germany than those which are limited in the east by a certain degree of winter cold. This climatal influence is especially seen in the valley of Göttingen, which does not possess those peculiar plants of Thuringia which find a warmer summer there, because the easterly winds are cooled down before they reach Göttingen, by the rough peaks of the Eichsfeld and Hartz ; while a few Thuringian species do occur sporadically in the Weser region at a distance from this plateau; as for instance, *Sisymbrium austriacum, Inula hirta, Melica ciliata,* &c.

Sect. 5.

The East-European Plain.

This, the largest of the European plains, is bounded on the east by the Ural Chain, stretching from N. to S., and by the Caspian Sea ; between these the flat country passes, interrupted only by the river Ural, into the plains of northern Asia. The southern boundary is formed by the Caucasus, extending from E. to W., the Black Sea and the Balkan ; on the west lie the Carpathians, but to the north of them the plain joins the North-European ; the Baltic is a natural limit on the west as far as Finland; while on the north lie the Frozen Ocean and its bay, the White Sea. This vast tract of land lies between the 43° and 27° N.L., thus extending through 27 degrees. From north to south it measures 400 geographical miles ; from west to east 200 between the Gulf of Finland and the Ural Chain ; 300 between this and Grodno, and 250 between the Carpathians and Astracan on the Caspian.

A small portion only of its borders is washed by the sea, namely, only where the Frozen Ocean and the White Sea, the Baltic and the Gulf of Finland, and the Black Sea, with the Sea of Azof, bound it. Much the greater part adjoins the main-land, and excepting at the Frozen Ocean, this plain is far removed from the great oceans of the globe.

This immense region is not altogether flat; elevations of the surface occur, especially in the N.W., but these are mere undulations, forming only ridges or high banks of rivers and seldom exceeding 800 or 1000 feet; they do not deserve the name of mountains or mountain-chains; such are the so-called Waldai and Volchonsky mountains, from which flow the Wolga and Dnieper towards the E. and S., and the Duna and Wolkow to the N.W. and N. The watershed between the Dnieper on one side, and the Duna and Niemen on the other, con-sists of morasses. These three rivers are united by a canal, as is also the case with the Dnieper and Weichsel; violent rain-storms also effect an occasional junction be-tween these oppositely flowing streams.

The north-western portion of this plain is the most elevated, sinking towards the N. and E. to the White and Caspian Seas. The level of the latter lies below that of the ocean, according to the Russian measurements about 83 feet, according to Hommaire de Hell, 60 feet below that of the Black Sea, and a very large tract of country surrounding the inland sea on the W., N. and E. forms a basin, which also lies below the ocean level. Saratow and Orenburg, 70 geographical miles in differ-ent directions from the Caspian, are level with the ocean, and Burzak by Lake Aral is below it. From hence the land rises to the N.W.; Kasan is 120 feet, Moscow 480 feet, and Gaisekain in Livonia, in the vicinity of the Baltic, rises to a height of 1825 feet.

The rivers of this plain are, it is well-known, the largest in Europe, and mostly arise within it. The northern part contains also larger lakes than any other part of Europe ; among which may be named Ladoga, Onega, and Peipus. Toward the south and east lakes are rarer. In the western portion bogs are met with.

The following table will afford some idea of the temperatures :—

	Elev.	Lat.	Ann.	Winter.	Summer.
St. Petersburgh	..	60°	$38\frac{1}{2}$°	16°	62°
Moscow	484 ft.	56	$38\frac{1}{2}$	11	66
Kasan	120	56	$36\frac{1}{2}$	11	63

Comparing these data with those before given for western Europe, their great inferiority is seen. The annual mean of St. Petersburgh is $3\frac{1}{2}$° lower than that of Stockholm, and nearly 6° below that of Ullensvang. Moscow is nearly 9° below Copenhagen, and $9\frac{1}{2}$° below Edinburgh. The winters are particularly cold and the summers proportionably warm. The summer temperature of St. Petersburgh is as high as that of Stockholm or Ullensvang, but the winter temperature is 9° lower than that of Stockholm, 14° lower than that of Ullensvang. The winter of Moscow is more than 20° below that of Copenhagen, 28° below that of Edinburgh, and more severe even than that of the North Cape or the north coast of Iceland (see Map). On the other hand the summer is warmer than in Copenhagen or Edinburgh, and is only equalled in Paris and Carlsruhe in the Western Europe. Thus, while Copenhagen appears to have a continental climate in comparison with Edinburgh, it has a coast climate compared with that of Moscow. At Kasan, in the same latitude as Edinburgh, mercury is sometimes frozen, which occurs in Lapland in the west. The winter is also proportionately cold in the south of the plain; the Sea of Azof is frozen at its margins every year, and the same happens to the Wolga where it pours into the Caspian Sea.

With regard to the fall of rain, the data are very incomplete. In St. Petersburgh it amounts only to 21 inches, and the westerly winds are most prevalent, although not to the same extent as in western Europe; they are also predominant in Moscow and Kasan. In the southern steppes it is stated that the average of four years has given only 6 inches fall of rain, occurring in 47 days of the year; but the irregularity is so great, that single years gave 59, 35, 39 and 53 rainy days. In 1832–3, twenty months elapsed without rain, and in some years the quantity is only one-tenth of that which falls in wet years. In the summer there is no dew, and the ground dries up and cracks, the plants withering up. 1841, not considered as a dry year, gave only $8\frac{1}{2}$ inches of rain; but in 1831, one of the wettest, the moisture interfered with agriculture more than the drought does, saturating the soil, which rests on a deep impermeable clayey formation. The prevalent winds are the N.E., E. and N., or polar currents, more than doubling the S.W., W. and S., equatorial currents, the average of four years giving 234 days of the first and 105 of the second, while the other directions S.E. and N.W. only prevailed, on an average, 31 days. These observations were made in Askania Nova, in the Nogaisch Steppe lying between the Dnieper and the Sea of Azof in the Tauris.

The Arctic portion of Russia differs in character from Scandinavian Lapland, in the circumstance that the limit of the growth of forests recedes almost to the vicinity of the Arctic Circle, so that extensive low and treeless tracts spread along the Arctic Ocean. Pine forests are wholly wanting at Kanin, with the exception of a forest of spruce (*P. Abies*), now disappearing, situated below $67\frac{1}{4}°$ N.L.; they cease at the Indega river, about fifteen English miles from the sea, and scarcely extend over the Arctic Circle beyond Petschora. The cultivation of barley and potatoes is not carried on beyond the city of Mezen. The forests are succeeded northward first by a zone of low birches and willows, then by the dwarf birch and the arctic Ericaceæ; with the latter the continuous turf ceases, and a few Ranunculaceæ, Saxifrages and grasses, growing in isolated tufts, are all that are subsequently met with. In an investigation of the eastern part of Archangel, the peninsula of Kanin and the island of Kalujen, M. Ruprecht found only 342 flowering plants; but these contained a considerable number of species, belonging to the genera *Ranunculus, Viola, Parnassia, Salix, Poa*, and others, which are not found in Swedish Lapland.

To the south of the above regions begin the provinces of Northern Russia, which are characterized by dense forests, in which the Scotch and spruce firs predominate, and the vast extent of which is only interrupted by swamps, or where, in the vicinity of the rivers, they have been thinned or destroyed by man. Among the firs are intermingled here and there alders (*A. incana*) and birches (*B. pubescens*), which in places form by themselves large forests. The limits between cultivation and the wilderness are especially marked by abundance of alder bushes. The aspen, mountain ash, and bird-cherry (*Prunus Padus*) are the only other trees.

The pine forests are of two kinds, according to the soil; the clayey and often marshy lowlands upon the old red sandstone are covered by dense woods of the spruce, among which grow the aspen and alder; while the low, sandy diluvial hills bear the Scotch fir and the birch, and resemble more the forests of the North-German plain. On this diluvium, where the clay is deficient, are also found open heaths clothed with the heather (*Calluna*), which do not occur on the Silurian plains and on the trap. At the same time the diluvium is not free from bogs, and in these occur *Ledum* and *Andromeda calyculata*; the spruce, however, does not grow on them, but only the Scotch fir, which does not shun water, and only requires a light sandy soil.

The characteristic plants of the pine forests are here, *Rubus arcticus, saxatilis, Chamæmorus* and *Idæus, Vaccinium Myrtillus, uliginosum* and *Oxycoccus, Rosa canina* and *cinnamomea*, and *Linnæa borealis.* *Gnaphalium dioicum* and *Cetrariæ* are found in the pine and birch forests principally; the forest meadows are filled with *Ranunculus reptans.* On the mountain limestone grow *Habenaria albida* and *viridis*; and on Lake Onega, *Aconitum septentrionale* (Mart.) is very luxuriant.

Two vegetable formations have been distinguished in the bogs of the clayey lowlands of N. Russia, viz. 1. the Dwarf Birch formation, where the swamp, of uncertain depth, is covered by a close quaking turf of *Sphagnum*, with the cranberry, which is sprinkled all over with bushes of birch (*Betula nana* and *fruticosa*), rising from 3 to 5 feet above the surface. In common with these grow many of the heath tribe with northern *Rubi* and willows, as *Ledum palustre, Andromeda polifolia* and *calyculata, Arbutus Uva-ursi, Vaccinium Vitis-idæa* and *uliginosum, Rubus arcticus, Chamæmorus* and *saxatilis,* with *Salix bicolor, limosa, glauca, myrtilloides* and *ros-marinifolia.*—2. The formation of Sedges. Here the soil is covered with water, but the bottom is firmer and more clayey than among the birch bushes, and the bog-moss is absent. Tufts of sedges occur crowded over the surface, and above these project abundance of the white heads of the cotton-grasses. There are no woody plants here; but *Calla* and *Pedicularis* are plentiful in places, and the open pools and lakes which occur on these swamps present almost the same forms as in Scandinavia and Germany; such as the white and yellow water-lilies (*Nymphæa, Nuphar lutea* and *pumila*), *Stratiotes,* the frog-bit, the aquatic *Ranunculi,* and *Caltha.*

The vegetation of this region is evidently analogous to that of the lower tracts of Swedish Lapland, but being situated under colder isotherms, it occurs here over vast tracts of lowland. The cultivated spaces only form oases in these boundless plains, which extend from the White Sea as far as the watershed near the district of the Wolga, everywhere covered with the four formations above described. The country is intersected by the valleys of the rivers in a peculiar way, as broad water-courses forming deep ravines in the great plain, which elsewhere is only slightly undulating. Thus, Ustjug-Weliki, on the Dwina, is only 330 feet above the sea, and the highest plateau of the forest-plain in the neigbourhood is, in general, about 600 feet. The swamps are usually found, extending many miles, along the sides of the broad low ridges. Towards the rivers the plain is depressed suddenly, and forms two terraces below the forest; the upper terrace, from 40 to 60 feet above the bottom of the valley, is undulating, and in great part cultivated, exhibiting also sloping meadows, blooming with Orchidaceous, Labiate and Compositous plants, but giving place lower down to bogs; all the hollows, especially along the borders of the forest, are occupied by marshy meadows. The lower terrace is quite horizontal, and reached by the overflowing of the river. It is uninhabited, and contains fertile meadows, banks bare of vegetation, and islands. The water-course everywhere lies on the right side of the valley, close to the base of the steep upper terrace. On the desolate sandy banks, throughout Russia down to the southern steppes, *Salix acutifolia* grows with roots from 40 to 60 feet long, closely interlaced among the loose soil. The meadows are in the first instance produced by the deposition of clay and marl from the river, and being annually irrigated and supplied with a fresh deposit of marl, possess a most luxuriant turf. On Lake Onega the dunes bear only *Calluna* and *Empetrum*.

As far as the condition of the soil is concerned, the land is everywhere adapted for the cultivation of all the Central European cereals, but the climate is adverse to agriculture. The destruction of the forests, which has been so ruinous in Central Russia, has effected little alteration in the character of these regions, and that only in the neighbourhood of the river valleys; yet within the memory of man, two fine and most useful trees have almost entirely disappeared. In the regions where Pallas still found wide ranges of larch forest, M. Blasius, in 1840–1, saw scarcely half a dozen trees in a distance of sixty or eighty miles; and, in like manner, *Pinus Cembra*, the Siberian pine, or 'Russian cedar,' which formerly extended further to the west, is now first met with in central Witschegda, east of the Dwina. The finest forests M. Blasius saw were those along the course of the Suchosia in the government of Wologda. Here the stems of the fir and aspen attain from 100 to 150 feet, and the birch not unfrequently 100 feet.

Central Russia is defined by the line of the watershed of the north and south streams, at the ridge of the Waldai Hills. The level is not more than 200 feet above the highest points in the north, and may be stated to have an average elevation of only 800 feet above the sea; yet this low range everywhere distinctly divides two extensive botanical regions. It is the south limit of the hoary alder (*A. incana*, D.C.), and the northern limit of orchards and of many deciduous trees, as of *Betula corticifraga*, which is at first mixed with *B. pubescens*, but further south exclusively constitutes the birch woods. The pines and firs decrease, and the aspen becomes more abundant, forming dense forests. This and the birch contend for mastery with the pine until the oak appears, and from this point the deciduous trees predominate. The ash, the lime, and the oak (*Q. pedunculata*) first appear near Jareslaw (*Q. Robur* is exotic to Central Russia, and eastward does not appear even to reach the Dnieper). The underwood consists of the hazel, mixed occasionally with the spindle-trees (*Euonymus europæus* and *verrucosus*) and the buckthorns (*Rhamnus catharticus* and *Frangula*). Jareslaw is also the northern limit of the following plants, which probably belong to that section of the 'Germanic type' which have their northern limit defined pretty nearly by the parallels of latitude (see page 169): *Berteroa incana, Lunaria rediviva, Lavatera thuringiaca, Chærophyllum aromaticum, Eryngium planum, Scrophularia vernalis*, &c.

The northern marsh willows are replaced by *Salix fusca, cinerea* and *Capræa*, and *Alnus incana*, D.C., by *Alnus glutinosa*, G.   Thus all the vegetable formations assume a new character ; but the physiognomy of the country is much more strikingly altered by the increased amount of cultivation.   The extent of cleared and now cultivated land is equal to that of the forest in Central Russia.   On the Oka, where the woods are composed of oak mingled with aspen and birch, they are generally limited to the vicinity of the rivers and the adjacent valleys and ravines ; the open country indicating the gradual commencement of the treeless steppes.   Here, on dry elevations, is already seen a dense vegetation of wormwoods (*A. scoparia*, Kit., *vulgaris, campestris* and *Absinthium*), which extends to the willows on the banks of the rivers, where *S. acutifolia* now grows intermixed with new species, such as *S. alba, fragilis, viminalis*, &c. Central Russia is geologically defined on the north by the predominance of magnesian limestone over the old red sandstone of Northern Russia ; further south its natural character is marked by the marly soil of the new red sandstone and mountain limestone, or by the calcareous marl, which present themselves tolerably free on the surface in long tracts.   On the Oscro, the central region of vegetation encroaches to some extent on the northern, owing to the calcareous soil ; on the other hand, between the Dwina and the Dnieper, the northern botanical region extends further south in consequence of the opposite geological conditions there existing.

Southern Russia commences where extensive diluvial deposits cover the limestone and tertiary formations, and are themselves again covered by the *black mould* or 'Tschernen Sem,' as it is called by the natives.   On the

Dnieper its northern limit lies near Tschernigoff, whence it passes through the southern part of the province of Kursk, and reaches the Wolga near Simbirsk, at which point the sandy covering of the limestone is immediately contiguous to the new red sandstone of the north. Thus the vegetation of the steppes is generally as distinctly defined from that of the region of the deciduous trees, as the latter from the region of the conifers. But Ukraine Proper, or the province of Kharkow, forms a peculiar transition between the steppes and Central Russia. It is a hilly country, the calcareous rock projecting through the diluvial sand. Hence forests occur covering a considerable portion of this fertile country, and on passing from the plain of Poltawa towards Kharkow, the ' black mould ' is observed to diminish in thickness on the watershed of the district of the Dnieper and the Don near Walki. Here the forests first appear; they consist of oak, lime, aspen, poplar, ash, and a maple (*Acer tataricum*), but always mixed with the wild pear; the underwood is chiefly hazel. The bare surfaces are here thickly covered with steppe-shrubs, 2 to 3 feet high, consisting of the leguminous plants *Cytisus supinus* and *Caragana*, with the dwarf cherry (*Prunus Chamæcerasus*, Jacq.). The flora of this province being decidedly S. Russian, seems to show that the climate exerts the greatest influence, since the soil of the Ukraine is generally calcareous, like that of Central Russia. Many plants are excluded by the low annual mean temperature, while the great summer heat appears favourable to the culture of maize and several Cucurbitaceous plants: it is said that the berries of the black nightshade (*Solanum nigrum*) lose their narcotic principle here, and when ripe become sweet and eatable.

The forests and fields of S. Russia are protected from the

continued droughts of the summer, which act so powerfully on the adjacent steppes, by the enormous quantity
of humus, the black mould, which is sometimes as much
as 10 or even 15 feet thick. Hence the nature of the
principal forest-trees, the oak, lime, elm, and wild pear,
these sending down deep roots : the spruce, which predominates in the scanty soil of the north, is unknown in
the Ukraine, and even the ash is sometimes killed by the
dry seasons. This deep soil causes several herbaceous
plants to grow to an unusual height : *Cephalaria tatarica*
is found 9 feet ; the tall larkspur (*Delphinium elatum*)
5–6 feet high ; and the thistles and umbellifers are generally twice the size of those of other regions ; the cap of
certain toadstools (*Polyporus* and *Leuzites*) is found
3 feet broad. But the most remarkable object is a kind
of puff-ball (*Lycoperdon horrendum*), spherical in form
and 3 feet in diameter. M. Czerniaïew says that it is
calculated to produce no slight terror, when the wanderer
comes in sight of it in the dark forest, where it looks like
a phantom in white or brown garments in a stoóping
attitude. The higher fungi generally are extraordinarily
numerous in the Ukraine ; more than 1000 of the Pileate
section being stated to exist there, while the abundance of
the *Gasteromycetes* is still more characteristic. The thick
rich soil is of course exceedingly favourable to agriculture,
altogether dispensing with the necessity of manures.

On the Desna, which falls into the Dnieper near Kiew,
first appear the wild fruit-trees, the wild pear, crab-apple
(*Pyrus communis* and *Malus*), together with the morello
cherry (*Prunus Cerasus*), and with them commences the
southern region of vegetation. These trees are distinguishable even at a distance from other deciduous trees,
by their crooked, crowded branches and dark-coloured

bark; the apple-trees have a trunk about 6 feet high, from which they send out a number of equal branches. But the general surface of the country is entirely without trees, and it is only in the swampy hollows and in the bottom of the river valleys (in the northern regions, the only parts which are clear) that an arboreal vegetation flourishes; but even here there are nowhere continuous woods, as far as the diluvial deposit extends over the surface. The conifers have long before disappeared, and the birch soon vanishes. The oak is the most abundant tree, and as in the Ukraine is always associated with fruit-trees; these form narrow strips of wood, very inconsiderable in proportion to the size of the steppe.

Cultivation is confined to the ' black mould,' on the border of the steppe. This narrow strip scarcely reaches to Krementschug, on the Dnieper, where M. Blasius found the northern limit of the culture of the vine. Close to this point the *Steppe* commences, with straggling herbs, intermingled with tall dry grass.

The Steppes of Russia, forming part of the vast plains extending from the borders of Hungary almost to China, constitute one of the most remarkable phænomena exhibited in the physical geography of Europe. Forming an almost uninterrupted plain, presenting merely slight undulations, rarely assuming the character of hills, the whole country lies slightly elevated above the sea, terminating with an abrupt terrace, rising 120 to 180 feet above the water, at the Black Sea, and sloping down gradually, in barren sandy tracts, often impregnated with salt, towards the Caspian. The rivers which intersect this plain, and which are swollen rapidly in spring on the melting of the accumulated snow, cut deep furrows in the surface, and, frequently changing their course, occa-

sionally leave dry ravines, slightly breaking the general
uniformity. The melted snow and rain-water flow away
but slowly from the level surface of the steppe; a portion is absorbed by the soil, but the greater part finds
its way gradually into the rivers, carrying away a sufficient quantity of earth to impart a black and turbid look
to all the streams that intersect the plain. None but
the principal rivers have any other supply than the rain
and snow, and thus are dry in summer. The elevation
being so nearly equal throughout the steppe, the fluviatile
ravines are of much the same depth everywhere, seldom
less than 100 feet, or more than 150 feet. Their sides
being abrupt, they form barriers to the herdsman and
traveller journeying across the plain, and in winter usually becoming filled with drifted snow, are then very
dangerous to those not well acquainted with the country.

The banks of the large rivers are less abrupt, but the
elevation, though gradual, is about the same; their beds
are generally very broad, and almost always fringed with
a belt of reeds 6 or 8 feet high. It is only in the vicinity
of the great rivers that the country assumes a diversified
aspect, and the wearied eye at last enjoys the pleasure
of encountering more limited horizons, a more verdant
vegetation, and a landscape more varied in its outlines.
" Among these rivers," says Mad. de Hell, " the Dnieper claims one of the foremost places, from the length of
its course, the volume of its waters, and the deep bed it
has excavated through the plains of Southern Russia."
Near Doutchina, on the post-road from Kherson to
Ikaterinoslaw, "it spreads out to the breadth of nearly a
league, and parts into a multitude of channels that wind
through forests of oaks, alders, poplars and aspens,
whose vigorous growth bespeaks the richness of a virgin

soil. The groups of islands capriciously breaking the surface of the waters have a melancholy beauty, and a primitive character scarcely to be seen except in those vast wildernesses where man has left no traces of his presence. These *plavniks* of the Dnieper have all the wild majesty of the forests of the New World."

The mouths of all the rivers flowing to the Black Sea open into large lakes, known by the name of *limans*; these are formed by the deposition of a bar across the mouth, in the tideless sea, from the vast quantity of mud brought down by the rivers. When the river is well supplied with water, the *peressip* or bar is kept open to the sea by a passage called the *gheerl*; but when this is not the case the communication becomes wholly cut off, and the liman then becomes a source of baneful exhalations, rising from the stagnant water under the hot sun of summer. Here and there over the plains, also, occur isolated shallow basins, called *stavoks*, which retain the moisture long after the rest of the country has been parched up by the summer heats, and these become at such time of no small importance to the herdsmen.

It has already been stated that the climate is one of extremes, and moreover very irregular. The intensity of the winter cold and of the summer heat are apparently but little influenced by the Black Sea, and the steppe, swept during two-thirds of the year by northern and eastern blasts, has an arctic winter and a tropical summer.

The depth of the long winter of the steppe occurs in December, January and February, during which the cold is intense, frequently surpassing the severest seasons known on the Baltic coast; but the winter is so stormy that it affords none of those advantages which it does in the North, where the hard, smooth surface of the frozen snow is everywhere cut into roads by sledges, and allows of locomotion only equalled in speed by railway-travelling. In the south the snow-storms are excessively violent, not merely consisting of heavy falls, but accompanied by winds that drift the snow and lift it in vast masses from the ground, filling the ravines, effacing the roads, and sweeping enormous quantities into the Black Sea, often covering it with ice for leagues from the shore.

The spring may be said to commence when the snow melts, which is usually in April, but May is sometimes far advanced before the mass of water has had time to find its way into the rivers. During this season the surface of the steppe is converted into a sea of mud, and a torrent of muddy water fills every ravine. The melting is often interrupted by a return of frost, and fresh falls of snow. For a few days, perhaps, a south wind will diversify the plain with tulips, crocuses and hyacinths, then a north-easter will come down from the Ural mountains, and cover everything with a snowy veil; and, with another change, a north-west gale will bring masses of heavy clouds, which fall and wash the whole face of the steppe from the Carpathians to the Ural. When the spring is fairly in, the steppe becomes clothed with a luxuriant herbage; but there are no trees to unfold their buds, nothing but grass and herbs, and these, though grateful enough to the eye after the frozen winter, are coarse and rank compared with those of Western Europe. Thunder and lightning are common throughout May, and the discharges are always accompanied by showers or heavy dews.

In June the thunder-storms cease, and the periodical

drought begins to approach. The soil dries up and cracks; the sun shines with burning heat, but the vapours raised from the soil are never returned even in the shape of mists; black lowering clouds sometimes sweep over the steppe, but hurry away to the Carpathians. Every trace of vegetation disappears except in some few favoured spots, near the great rivers; the surface of the soil becomes browner and browner, and at last almost black. Even on the cultivated tracts, on the 'black mould,' the summer is sometimes so dry as to destroy all the crops; and the corn, usually so luxuriant, then scarcely peeps from its furrows, with its red blades and grainless ears.

This is the more usual condition of the summers; but the seasons are extremely uncertain, and in some years the summer is very wet, thus interfering as much with agriculture. In such cases the wild vegetation of the steppe becomes very luxuriant.

In August the dryness reaches its extreme point, but before the end the dews recommence, and thunder-storms again occur, accompanied by rain. The sky once more becomes clear, and a fresh green herbage springs up over the plain; but the autumn is short, and October is a stormy month, accompanied by cold rain and fogs, usually closing with snow-storms.

The number of plants found upon the steppes is comparatively small, and a great many of them are social in their growth; thus in spring, tulips will be seen covering hundreds of acres; for leagues nothing will be seen but wormwood. The herbage varies in quality with the soil, and three classes have been distinguished according to the prevailing plants. The richest pasture is composed of *Festuca ovina*, with *Triticum cristatum* and *repens*;

and among these, herbs, such as lucerne (*Medicago fal-cata*) and thyme (*T. Marschallianus*), abound. The second-class pastures contain a greater quantity of the feather-grasses (*St. pennata* and *capillata*) overgrowing the *Festuca*; the herbaceous plants are rare, but there is still a good deal of *Triticum cristatum* and lucerne. The poorest pastures produce little else but feather-grass, and the few herbs occurring among this are useless for fodder.

The prevailing steppe-plants are: the grasses already mentioned, with sea-lavender (*Statice tatarica* and *lati-folia*), thyme, sages, wormwoods, sow-thistles, lucerne, spurges, pinks, mulleins and thistles; also in places, maritime plants like those occurring on sandy sea-shores.

In these regions the wormwoods and thistles grow to a size unknown in the west of Europe; it is said that the 'thistle bush,' found where these abound, is tall enough to hide a Cossack horseman. The natives call all these rank weeds, useless for pasture, *burian*, and, with the dry dung of the flocks, this constitutes all the fuel they possess. One curious plant of the thistle tribe has attracted the notice of most travellers, the 'wind-witch,' as it is called by the German colonists, or 'leap-the-field,' as the Russian name may be translated. It forms a large globular mass of light wiry branches interlaced together, and in autumn decays off at the root, the upper part drying up. It is then at the mercy of the autumn blast, and it is said that thousands of them may sometimes be seen coursing over the plain, rolling, dancing and leaping over the slight inequalities, often looking at a distance like a troop of wild horses. It is not uncommon for twenty or thirty to become entangled into a mass, and then roll away, as Mr. Kohl says, "like a huge giant in his seven-league boots." Thousands of them are annually blown into the Black Sea, and here, once in contact with water, in an instant lose the fantastic grace belonging to their dry, unsubstantial texture.

The steppes of the Caspian are remarkable for their desert condition, being covered by shifting sand, and often strongly impregnated with salt.

Bessarabia.

Bessarabia, bounded on the S. by the Danube, N. and E. by the Dnieper and Black Sea, and W. by the Pruth, which separates it from Moldavia, forms a long narrow strip 375 English miles long, with an average breadth of 50. Expanding as it approaches the sea, it is divided into two very distinct regions; the southern part, which the natives call Boudjak, is the flat country between the mouths of the Danube and the lower part of the Dniester. It has all the characters of the Russian steppe. The northern part, adjoining the Austrian possessions, is a hilly country, beautifully diversified, covered with magnificent forests, and rich in all the products of the most favoured temperate climates.

The Crimea.

This peninsula of Southern Russia is connected by a narrow isthmus to the great plain, and its northern part is flat, partaking of the character of the adjoining mainland; but on the south rises a small but comparatively high range of limestone mountains, very precipitous on the south face, and sinking gradually into the plain on the north. The highest points are Tschadyr-dagh, 4700 feet, and Babugan-Jaila, about equal in height.

Between these a cross valley runs with a pass, at an

elevation of 2000 feet.   The south coast of the Crimea is quite a distinct country both in climate and vegetation. The tract lying between the steep side of the mountains and the low sandy sea-shore is peculiarly circumstanced. It is entirely protected from all the rough winds of the north, while it is quite open to all the warm breezes blowing from the southern shores of the Black Sea.   It consequently enjoys an exceedingly mild climate, which allows the vine, olive, laurel, pomegranate, and all the South-European fruits to come to perfection.   The principal culture is that of the vine, but the wines are far inferior to those of Western Europe ; the vineyards extend at present into a number of the northern valleys of the Tauric chain ; but the wine is much poorer than that of the south coast.   Fruit is however largely cultivated in the northern valleys, with an exclusive privilege of supplying Moscow and St. Petersburgh.

The southern declivity of the ridge is covered with forests of the Corsican pine (*P. Laricio*), the region extending from 600 to 3000 feet.   On the north side, which is much colder, they are replaced by beech.   *Arbutus Andrachne* occurs only on the south side, from the coast up to 1200 feet, but usually solitary, and is supposed to have been derived from seeds brought by birds of passage from Anatolia.

SECT. 6.

The Central European Highlands.

BETWEEN the N. European plain on the one side and the Alps and Pyrenees on the other, the mountain masses of Central Europe form a semicircle round the highlands of the south of Europe. The breadth of this zone is greatest to the north of the Alps and diminishes to the east and west.

Eastward of the low western coast of France rise the mountains of Auvergne, forming a longish range, running about 20 geographical miles from N. to S., between the rivers Allier, Dordogne and Lot. The mountains are flattened at the top so as to form elevated plateaux (3500 feet), but from these rise much higher points, mostly extinct volcanos of conical form. The highest of them are Mont d'Or, 6177 feet, and Cantal, 6075 feet.

The Cevennes form a long narrow ridge on the east and south of the Auvergnese mountains, extending first from S.W. to N.E., then from S. to N., their length amounting to 45 geographical miles. They are separated from the Pyrenees by a deep depression, in which lies the Canal of Languedoc, connecting the Mediterranean with the Atlantic, at a level of about 500 or 600 feet above the sea. Their eastern boundary is the Rhone, their western the Tarn and the Allier. The northern part is known by the name of the Montagne Forez. The mean height of the Cevennes is about 3500 feet ; the highest summits, Pierre sur Haut, 6336 feet, Mont Mezin, 5654 feet, and La Lozère, 5300 feet.

The Jura, also an elongated mass of mountains, stretches from S.W. to N.E., between the rivers Saone and Doubs on one side, and the north-west of Switzerland and the Rhine on the other. The Swabian Jura, on the other side of the Rhine, extending in the same direction and composed of the same limestone, may be regarded as a prolongation of the Jura. The length of the Jura Proper is 35 geographical miles, with the Swabian Jura 60 ; it is composed of several elongated chains,

P

between which lie large, flat, long valleys, with very few cross valleys. The mountains sink precipitously towards Switzerland, in terraces towards France; the mean height is about 3500 feet. The highest points are almost on a level, namely, Pré de Marmiers 5631 feet, Réculet 5610 feet, Mont Tendre 5503 feet, Dôle 5499 feet. The Swabian Jura, between the Neckar and the Danube, averages about 1600 feet; Hohenberg, the highest summit, reaches 3357 feet.

On either side of the Rhine, and on the north of the Jura, lie several ranges of mountains, which may be termed the Rhenish Mountains. On the west of the Rhine lie the Vosges, a largish mass, running from S. to N. for about 30 geographical miles, bounded on the west and north by the Moselle, with a mean height of 2500 feet, and dome-shaped peaks, among which the Ballon de Sulz, 4608 feet, and the Ballon d'Alsace, 4111 feet, are the highest. On the east side of the Rhine, running parallel with the Vosges, is the Black Forest, shut in by the Neckar, and the Odenwald, surrounded by the Maine on the E. and N. The former mountains have an average of 2500 feet, with one summit, the Feldberg, of 4947 feet; the latter rise only to 1000 feet, the highest point, Malchen, only 1671 feet.

Lower than these and more detached are the northern Rhenish Mountains,—Spessart, Taunus, Vogelgebirge and Westerwald, on the east of the river; the Hunds-rück (S. of the Moselle) and the Eifel (N. of the Mo-selle), and others on the west side.

The name of the Weser group has been applied to a number of ranges from which the waters feed that river.

The Hartz is a detached chain, flat on the top, 2000 feet above the sea, with a number of peaks rising out of this, among which the Brocken, 3725 feet, is the highest. The proper Weser Mountains are much lower, not averaging more than 600 feet, the highest point being 1500 feet. The Rhöngebirge and several basaltic ridges are also low. The Thüringerwald and Frankenwald (the highest point Great Beerberg, 3366 feet) form the transition to the chains in which the Elbe takes its origin.

The central portion of Bohemia, particularly on the north side, resembles a basin, surrounded by mountains which may be collectively called the Elbe Mountains; they are for the most part flat on the top, and thus approach the form of ridges; but higher summits rise above them. These masses, enclosing the Bohemian basin, are: the Fichtelgebirge on the N.W., 1600 feet (highest point Schneeberg, 3423 feet); the Erzgebirge on the N., 1600 feet, falling abruptly into Bohemia, gradually towards Saxony (Schwarzwald, 4000 feet); the Böhmerwaldgebirge on the W. and S., an elevated plateau of 2500 feet, above which rise Heidelberg, 4100 feet, and Arber, 4080 feet. The Sudeten, the northern part of which is known as the Riesengebirge, on the N.E. averaging 2500 feet, in which Schneekopf, 5255 feet, and the Glatzer Schneeberg, 4568 feet, are the highest summits.

The Hungarian plain resembles the Bohemian, but forms a much larger basin. On the W. and S.W. it is bounded by the most eastern portion of the Alps; on the S. by the Dinaric Alps; but two-thirds of the circumference of this almost circular plain are bounded by a flattened dyke, as it may be called, extending about 140 geographical miles, with an average height of from 2000 to 3000 feet. Upon this elevated tract lie several groups of mountains, some of considerable height. The only openings into the Hungarian plain are at the gaps where the Danube enters and emerges. The great surrounding wall and its groups of mountains constitute the Carpathians. The loftiest of the groups, Tatra, lies in the north-western part of the range, and the highest peaks in this group are the Eisthalerspitze, 8500 feet; Lomnitzerspitze, 8440 feet; and Hundsdorfferspitze, 8288 feet. The southern portions of the Carpathians also attain a remarkable height.

Between and beside these elevated masses lie many plains and valleys of considerable importance. To the west of the Auvergne Mountains and the Cevennes, the country spreads out into the level, sandy tract called the *Landes*. Between the Cevennes and the Côte-d'Or on one side, and the Alps and Jura on the other, lies the valley of the Saone and Rhone. The watershed between the Mediterranean and the North Sea rises to 1400 feet, and they are connected by a canal here. The valley of the Rhine, between the Vosges on the west and the Black Forest on the east, sinks from 800 feet down to a low level. Between the Swabian Jura, the Black Forest, the Odenwald, the Thüringerwald, the Fichtelgebirge, and the Böhmerwald, lies the plain of Franconia, at an elevation of 900 feet; the Bohemian basin is 500 feet above the sea; the Moravian plain, between the Sudeten and the Carpathians, 600 feet, passes gradually into the eastern part of the chains of the Böhmerwald, and thus indirectly into the Bohemian depression. Between the Böhmerwald and the Alps lies the great valley of the Danube or Lower Austrian plain, and this is immediately continuous with the vast Hungarian plain, which is elevated only 200 to 300 feet above the sea.

The rivers rising among or running between the central European mountains are very numerous, and of great importance. From the variety of directions in which they flow, their long and often rather tortuous courses, together with the frequency of rapid currents, they must undoubtedly have had much influence in producing the present condition of the distribution of the vegetation of Central Europe. The most considerable are—the Garonne, with its sources in the Pyrenees, Cevennes and Auvergnese mountains, running into the Atlantic; the Loire and Seine, deriving their waters partly from the two last groups and partly from the northern hills, flowing, the first directly into the Atlantic, the second into the Channel; the Rhone, flowing into the Mediterranean from the Cevennes, the eastern face of the Jura and the west side of the Alps; the Rhine, arising in the Alps, receiving tributaries from the Jura, Vosges, Black Forest, Odenwald, and many smaller elevations; the Weser, from the Weser Mountains and the Hartz; the Elbe, arising in the mountains surrounding the Bohemian basin, all these thence pouring their waters into the North Sea. The Oder, springing from the eastern declivities of the Sudeten, and the Weichsel from the north side of the Carpathians, running into the Baltic. The Dniester, flowing from the east side of the Carpathians into the Black Sea; and lastly, the Danube, the sources of which lie in the Black Forest and Swabian Jura, and which, in its course towards the Black Sea, gathers the contributions of large tributaries from the E. and N. sides of the Alps, from the mountains of the Böhmerwald and from the Carpathians.

The only large lakes are in Hungary, namely the Platten-See and Neusiedler-See.

The following table gives a summary view of the temperatures :—

	Lat.	Ann.	Wint.	Sum.
Bordeaux	45°	57°	43½°	71°
Carlsruhe	49	51	34½	66
Prague	50	50	31	67
Vienna	48	50	32	68
Ofen (500 ft.)	47½	51	31	70
Clermont, (Auvergne, 1344 ft.)	46	51	36½	64

The mean annual temperature also decreases here towards the east, while the summers become hotter and winters more severe. The valley of the Rhine (Carlsruhe) and the Bohemin basin (Prague) have a comparatively high mean. The winter of Prague (50° N. L.) equals that of Copenhagen (51½° N. L.), but the summer is 3 degrees warmer. In Vienna and Ofen the difference between winter and summer amounts to 36°, in Paris to 26½°, in Bordeaux to nearly 28°. Clermont, situated on the plateau of Auvergne, 1260 feet above the level of the sea, has the same mean temperature as Paris, lying 3 degrees further north. Compared with Bordeaux, the mean temperature of Clermont gives a decrease of 6 degrees, and this supposes a diminution of one degree of Fahrenheit for about every 220 feet.

The annual quantity of rain amounts to about 24 inches in the W. of France, that is, rather less than in Holland and the British Islands ; but in the vicinity of the mountain-chains, and in the valleys between them, for instance, in the Rhone and Rhine valleys, it becomes considerably greater ; towards the east it again diminishes very much ; it is only 15½ inches in Prague, 18 inches in Ofen.

Although the Carpathians rise so high as 8000 feet, they do not quite reach the snow-line. On the Eisthalerspitze only a little snow remains in the ravines through the summer months, and there is a small glacier there. The rest of the central European mountains, none of which exceed 6000 feet, are all below the line of eternal snow. But the higher peaks of the Carpathians, the Riesengebirge, the Jura, Cevennes, and Auvergne mountains are covered with snow during the greater part of the year, as their names, Schneekoppe (Snow-top), Schneeberg (Snow-mountain), Schneekopf (Snow-head), and the like, indicate.

The western coast of France is chiefly occupied by heaths (*Landes*), but coniferous woods exist, composed of the Aleppo or coast pine (*Pinus halepensis, maritima*). The mountain masses of France, the Jura, Cevennes, Auvergne, are chiefly wooded with beeches and oaks in the higher regions, in the lower with the chestnut. In the German mountains the forests are partly composed of Scotch fir (*P. sylvestris*), Norway spruce (*Abies excelsa*) and silver fir (*P. Picea*), hence the names of the Black Forest, and the Fichtel-gebirge (*fichte* means the spruce fir), and partly of beech and oak. The chestnut also occurs in the valley of the Rhine, and in the side valleys of the Maine and Neckar. In the Carpathians the dwarf pine (*P. Mughus*) and the Norway spruce form the elevated woods; the beech predominates below.

In the great extent of country included within the present region, there exist, of course, very considerable diversities in the character of the vegetation, but, taken as a whole, it may be regarded as the especial region of the Central European Flora. On all its borders it is invaded by plants belonging to contiguous districts; and, passing from west to east, from the maritime climate of western France to the inland plains of Hungary, there are naturally found great differences, depending upon the altered climatal conditions. Yet the *mass* of the flora is nearly the same throughout.

The climate of the west of France is characterized by mild winters, and summers tempered by the influence of the ocean; and as many southern plants can accommodate themselves to such conditions, we find numbers of them advancing up the coast as far as the Loire, such as *Quercus Ilex* and *coccifera, Iris tuberosa, Amaryllis lutea*, &c.; others even as far as the mouth of the Seine, as *Astragalus monspessulanus,* &c., but the greater part of them do not spread out of the basin of the Gironde. These plants, however, are so numerous as to give the flora of the plains of the S.W. of France quite a distinct character from that of Central Germany, so that our definition of the vegetable regions would be false here, were it not for the great development of the highlands in this part of the district, which strongly characterize it, and manifest the closest relations to the highlands of central Europe. The south-east corner of France, on the south side of the Cevennes, belongs wholly to the Mediterranean region, its characters allying it with Spain and Italy rather than with central Europe. The limit of the Mediterranean flora here lies about at the watershed between that sea and the Atlantic, in the province of Languedoc. Of this we shall speak hereafter.

The flora of the south of France, excluding the Mediterranean district, is composed of about 750 species, in which the most numerous families are the Leguminosæ, the Compositæ, the Grasses, Cruciferæ, Umbelliferæ, Labiatæ, Caryophyllaceæ, Liliaceæ, Cyperaceæ, &c., in the order enumerated. The richest genera are *Centaurea, Trifolium, Carex, Medicago, Helianthemum, Lathy-*

rus, Vicia, Galium, Sedum, Linaria, &c., indicating that the Dicotyledons are becoming relatively more numerous than the Monocotyledons, which is true not only of the species, but of individual plants. The southern character is marked by the grasses and sedges becoming less predominant, while the Leguminosæ, Compositæ, Cruciferæ, Umbelliferæ and Labiatæ gain in importance.

The mountains of Auvergne present several zones of vegetation. At about 3000 feet is reached the mountain plateau, on which rye is the cultivated grain. The Scotch fir is the prevailing tree, the beech grows upon the ancient lava streams, *Salix pentandra* along the brooks, and the ash in the enclosures. The presence of *Actæa spicata, Lilium Martagon, Gentiana cruciata* and *asclepiadea,* and *Salvia Sclarea,* at once indicate the elevation above the sea and the relationship to the mountains of the Upper Rhine. Passing from the plateau to the Mont d'Or, are found rye, barley, oats, and even summer wheat ; hemp succeeds up to 3300 feet, but the vast mountain pastures principally distinguish this region, in which are found *Trifolium alpinum, Armeria maritima, Gentiana lutea,* while *Ribes petræum, Lonicera alpigena, Mespilus,* the mountain ash, and other subalpine plants grow vigorously among the rocks. The pasture is principally composed of *Nardus stricta,* with alpine grasses. The silver fir (*A. pectinata*) ascends to about 4000 feet. On the summit of the Puy-de-Dome, 4900 feet, *Ajuga alpina, Trollius europæus, Athamanta Libanotis,* and other subalpine plants occur ; but on the Puy-de-Saucy, 6322 feet, *Soldanella alpina, Saxifraga cæspitosa, Gentiana nivalis, Euphrasia minima,* and *Cerastium alpinum* have been gathered, all true alpine plants. The rhododendrons do not occur on the mountains of Auvergne, the only alpine shrub being the dwarf juniper.

The vegetation of the Jura and the neighbouring mountains has been the subject of an elaborate work by M. Thurmann, from whom we derive the following characters. He divides it into four regions, according to height, characterized in the following manner.

1. The *lower* region or sub-Jurassic border, extending to about 1300 feet. The vine is met with here in tolerably exposed places; maize extensively in the S.W. districts. All the cereals are abundantly cultivated, as are also the fruit-trees. The walnut is tolerably plentiful; the oak widely distributed, forming forests, *Q. sessiliflora* often predominant, but less so in the Swiss basin. The beech also occurs over considerable extent, in forests. The spruce fir is not met with, nor *Pinus Picea*, excepting in the Swiss basin. All the species of the succeeding region are abundant and flourishing here when the soil permits, but most frequently occur sparingly from want of proper stations. The following is a list of 24 plants, one half of which, at least, may be found in every district of this region : *Stellaria Holostea, Hypericum pulchrum, Spartium Scoparium, Melilotus officinalis, Trifolium fragiferum, Ononis spinosa, Orobus tuberosus, Cerasus padus, Castanea vulgaris, Eryngium campestre, Pulicaria vulgaris, Senecio aquaticus, Onopordum Acanthium, Centaurea Calcitrapa, Hieracium boreale, Verbascum Blattaria, Stachys germanica, Quercus sessiliflora, Betula alba, Luzula albida, Carex brizoides, Aira flexuosa, Holcus mollis,* and *Triodia decumbens.*

2. *Middle* region of the Jura, from 1300 to 2300 feet. Here the vine is very rare, or altogether wanting; maize tolerably common in the western districts; all the cereals pretty common. Fruit-trees moderately common or scattered. The walnut pretty frequent, as also the oak, especially *Q. pedunculata*, forming forests. The

beech forests very common; the spruce fir sparingly, forming forests in the eastern Jura; *P. Picea* wanting or very rare; with absence or general rarity of the species of the lower region. Characteristic species:—*Helleborus fœtidus, Prunella grandiflora, Orchis pyramidalis, Orchis militaris, Fagus sylvatica, Euphorbia amygdaloides, Orobus vernus, Cephalanthera rubra, Bupleurum falcatum, Melittis Melissophyllum, Veronica prostrata, Melica ciliata, Buxus sempervirens, Sambucus racemosa, Euphorbia verrucosa, Convallaria multiflora, Coronilla Emerus, Aronia rotundifolia, Myosotis sylvatica, Calamintha officinalis, Carex alba, Anthericum ramosum, Teucrium Chamædrys, Daphne Laureola.* Most of these species run into the mountainous region, while some, characteristic of the latter, descend sparingly into the upper part of this middle region.

3. *Mountainous* region, from 2300 to 4300 feet. Without vines or maize; wheat occurring sparingly, barley and oats widely; the cereals ceasing, however, at about 3600 feet, and the fruit-trees, which occur less frequently, becoming rare or disappearing about 3300 feet. The walnut is not met with, and the oak is rare, seldom forming forests; the beech is tolerably common, mingled with the spruce fir, less frequently forming forests by itself. The spruce is now abundant, constituting pine forests, and *P. Piceà* changes from a rare to an abundant forest tree. The species characteristic of the middle region become much less evident towards 3300 feet; the most prevalent here are:—*Gentiana lutea, Trollius europœus, Crocus vernus, Rhamnus alpinus, Carduus defloratus, Abies excelsa, Mœhringia muscosa, Campanula pusilla, Arabis alpina, Ranunculus aconitifolius, Spiræa Aruncus, Lonicera alpigena, Geranium*

sylvaticum, Draba aizoides, Lunaria rediviva, Coronilla vaginalis, Athamanta cretensis, Saxifraga aizoon, Chærophyllum hirsutum, Bellidiastrum Michelii, Adenostyles albifrons, Centaurea montana, Abies pectinata, Prenanthes purpurea. Some characteristic species of the alpestral region descend here and there into the upper parts of this, most of the mountainous species ascend into the alpestral region.

4. The *alpestral* region, from 4300 to 5600 feet, is wholly uncultivated; both the oak and beech are rare, or absent throughout; the spruce is tolerably common, but rarely forms forests, while the *P. Picea* is abundant, constituting forests, but ceases about 4600 feet. Those species of the preceding region which inhabit forests cease with the latter, but the rest mostly persist. The greater part of those which characterize the middle region have disappeared. Prevailing species:—*Alchemilla alpina, Poa alpina, Potentilla aurea, Heracleum alpinum, Anemone narcissiflora, Dryas octopetala, Bupleurum ranunculoides, Hieracium villosum, Gentiana acaulis, Anemone alpina, Androsace lactea, Saxifraga rotundifolia, Sorbus Chamæmespilus, Polygonum viviparum, Helianthemum œlandicum, Gymnadenia albida, Ranunculus alpestris, Erigeron alpinum, Rumex arifolius, Sonchus alpinus, Nigritella angustifolia, Carex sempervirens, Phleum alpinum, Aster alpinus.*

The principal plains that surround the Jura are: the valley of the Rhine, the Swiss basin, and the valley of the Saone. The elevation of these seldom exceeds 1300–1600 feet, and the French portions are all below the first of these limits. Their vegetation is everywhere contrasted with that of the Jura, from difference of elevation, soil and exposure. The cultivation of the vine follows the whole of this region, and surrounds the base of the Jura almost completely, penetrating even into some of its valleys. These vineyards are connected to a certain extent with those of the adjacent low countries of France, Switzerland and Germany, but as in this latitude the vine flourishes better on declivities than on flat surfaces, there often exist interruptions between these vineyards of the sub-Jurassic zone and those of the surrounding countries, due to the presence of low, flat plains. The French side possesses many advantages for the culture of the vine, since the mountain slopes descend to a lower level there than on the Swiss side; and although the greater declivity of the former has a general exposure to the W. and N.W., its irregular promontories and valleys present innumerable slopes facing to the S. and S.W., while the soil also appears more favourable to the growth of this plant. Hence the superiority of the French wines. This, however, only refers to the vineyards situated south of Besançon.

The depression which separates the Jura from the Vosges, between Villersexel and Béfort, contains a few vineyards on the first half of its length, but they are absent on the other points. Most of the slopes face to the north; those which face the south belong properly to the Vosges, and are wooded. In Alsace they are scattered on the more favourable southern slopes; the valley of the Rhine from Basle to Schäffhausen exhibits an interrupted line of vineyards, where the products are inferior; from Schäffhausen to Constance, the elevation rising, the culture of the vine is always above 1300 feet, and runs to 1600 feet, the product being of very middling quality.

The vegetation of the valley of the Rhine presents

two regions, the Rhenish plain and the upper plain; the former, extending along both shores with a variable width of from three to nine miles, exhibits a great variety of conditions; sometimes the sandy banks are covered with willows, poplars, *Hippophae, Myricaria,* &c.; or stagnant marshes occur, abounding in aquatic forms, such as *Potamogeton, Typha, Scirpus, Butomus, Utricularia, Limosella, Œnanthe, Sium,* &c.; or vast tracts of meadow, frequently inundated, swarm with species of *Carex, Scirpus, Cyperus, Juncus, Pedicularis, Bidens, Sanguisorba, Asparagus,* &c., while in their drier portions many of the species of the middle region of the Jura are found; elsewhere the lower plain is covered with extensive forests, in which oak, hornbeam, alder and ash predominate, while the drier stations bear the birch, broom, and almost all the plants of the middle Jurassic region; lastly, sandy heaths appear with brooms and heaths, and all kinds of herbaceous sand plants. The *upper plain,* where the soil is less sandy, is more elevated, and is occupied by rich cultivation, or more verdant and luxuriant pastures, analogous to the marly meadows of the Jura, and containing few sand-loving plants. These characters apply most especially to the Alsatian district, but with slight modifications hold for the Bavarian bank and the valley of the Rhine generally.

The Swiss basin is bounded on the one side by the Lake of Constance and the continuous line of the high chains of the Jura, and on the other by the broken and irregular line formed by the advance of the first groups of the lower Alps. Its space is occupied by interrupted plains and numerous hills. The vine and finer cultivated plants form a border at the foot of the Jura by Zürich, Neufchatel, Lausanne and Geneva, and this region presents the greatest semblance to the lower region on the other side of the Jura. The principal stations of the sand plants analogous to those of the Rhenish countries occur on the plains of Eglisau, and around the lakes of Bienne, Neufchatel and Morat, as also in the basin of Leman, and here and there along the rivers. But the principal feature of the Swiss valley is the presence of forests of *Pinus Picea*. This tree, which is abundant in the Alps, descends by the easy gradations which unite the lower Alps to the Swiss hills, and finding there a favourable soil, it constitutes forests, either alone or associated with the beech, more rarely with the spruce fir or oak.

The geological structure of the Vosges differing so greatly from that of the Jura, it is not surprising to find a very marked distinction in the characters of their floras. The calcareous rocks of the latter here give place to crystalline and sandy formations, the granite *ballons* or dome-shaped hills covered with firs, the sandstone hills with beech, among which oak and birch are scattered; and in passing from the last Jurassic calcareous hills to those of the Vosgesian sandstone, in the zone between Béfort and Lure, a striking contrast is seen. The broom, the birch, *Luzula, Aira, Jasione*, &c., appear almost immediately, and in advancing, the oppositions become still more striking. To those plants which form the common basis of the vegetation of the country, a number of species are soon added which never occur in the Jura, or if but sparingly there become now widely spread; and as the Jurassic species become circumscribed, scattered or lost, the others take possession of the country. *Orobus tuberosus, Senecio sylvaticus, Luzula multiflora* and *albida, Spartium, Betula, Aira*

cæspitosa and *flexuosa,Vaccinium Myrtillus,* &c., become
frequent and common in the woods; the dry or moist
pastures become overgrown with heather, *Genista germanica, Agrostis vulgaris, Alopecurus pratensis, Juncus
uliginosus, squarrosus* and *Tenageya,* &c.; the sandy
tracts are decked with *Jasione montana, Ononis spinosa,
Filago minima,* &c. *Montia* becomes common in the
brooks; while the evident tendency of these and many
others to grow socially and form extensive patches, as
is the case with the brooms, heath, ferns, and some
grasses, gives the surface-vegetation a totally different
physiognomy from that of the Jura.

All these characteristics are heightened in the mountainous region by the increase of the species belonging
to moist and wet stations. *Vaccinium, Aria* and *Montia* become more abundant; to the *Junci* above-named
is added *J. filiformis,* with *Blechnum boreale, Lastræa
Oreopteris,* &c., *Luzula maxima, Genista pilosa, Nardus stricta, Meum athamanticum,* &c.; while a new
group, entirely strange to the calcareous chains, makes
its appearance, composed of such plants as *Arnica montana, Viola lutea, Silene rupestris, Sedum annuum,
Galium saxatile, Digitalis purpurea,* &c., giving a very
distinctive aspect. Forests of spruce, mingled with a
few birches, carpeted with *Vaccinium,* and shading
Digitalis and *Meum;* pastures spotted with *Arnica,*
turfed over large tracts with the rough *Nardus,* and
presenting an especial abundance of *Scirpus, Schœnus,
Carex* and *Juncus,* particularly *squarrosus,* in the damper
places; rocks always less bare, clothed with *Silene, Sedum* or *Galium,* and bearing elegant tufts of *Poa sudetica*
and *Calamagrostis sylvatica;* everywhere an abundance
of ferns, Lycopodiaceæ, mosses, and, above all, lichens,

coating the more exposed rocks; such are the general features which strike the observer in the Vosges, as contrasting with the vegetation of the Jura.

The subalpine region offers quite as striking differences. *Sedum repens, Saxifraga stellaris, Angelica pyrenaica, Leontodon pyrenaicum, Hieracium alpinum, albidum, Mougeotii, Androsace carnea, Luzula spadicea, Carex frigida, Poa supina, Allosorus crispus, Lycopodium alpinum* and several others, wanting in the Jura, give a peculiar character to the subalpine summits of the Vosges, approximating to that of the granitic Alps.

It should be observed also, that the subalpine species common to the Vosges and the Jura occur lower down in the former than in the Jura; and of the mountain and sub-alpine plants which occur in the Vosges and the Alps, the greater proportion do not appear nearly so low in the latter as in the former. Thus *Saxifraga stellaris* and *Leontodon pyrenaicum* occur in the Vosges at about 3600 feet, but in the Swiss Alps scarcely below 4300 to 4600 feet. *Hieracium albidum* and *alpinum* are found in the Vosges at about 4300 feet, but in the Alps inhabit an elevation of more than 5300 feet, and so on; which is accounted for by the inferior temperature of the Vosges at equal levels with the Alps, resulting from the more northern position of the former.

The vegetation of the Black Forest resembles that of the Vosges in its general characters. The most important of the differences depend on diminished temperature and increased humidity, so that still fewer of the plants of dry stations, characteristic of the Jura, occur here. This is strongly marked also by the descent of the spruce into lower regions than in the Vosges, and by the pre-

Q

sence of extensive forests of *Pinus Picea* which clothe this chain, and to which it probably owes its name.

The Albe or Swabian Jura, on the other hand, partakes of the character of the Jura, the resemblance being chiefly interfered with by the presence of sandy tracts on the plateau, and the interspersion of more Germanic species.

It would occupy far too much space to enter into a minute account of the districts of Central Germany, and we must therefore be content with selecting one or two provinces as representatives of the rest, and for an example of the middle region we may take the vegetation of the district around Jena. This district is an elevated plain, stretching out from the Thuringer-Wald in gentle undulations, having a maximum elevation of 1550 feet, and as the highest points only surpass the general level by 300 or 400 feet, it may be considered as belonging to the lower mountain region of Central Germany. In its geological structure it forms part of the stratified system, principally composed of rocks belonging to the red sandstone and Muschelkalk formations, with a small expansion of Keuper, brown coal, sandstone, alluvium or freshwater limestone of the tertiary system.

The vicinity of the Thuringian mountains renders the climate generally colder than in some places beyond the north-eastern limits of this region, so that the blossoming of fruit-trees and the harvest are a week or a fortnight earlier at Halle or Leipsic than at Jena. But in the deep and narrow valleys of the Saal and the neighbouring side-valleys, the summer temperature rises higher than in the places just named, and indeed sometimes attains an almost incredible height, considering the latitude. The prevalence of southern species in the flora of Jena proves the existence of a warm, favourable climate; and the almond and peach, with the quince, stand the winter on some of the vine-hills around Jena, and the vine itself flourishes excellently on the slopes of the Saal valley. Woods abound, the deciduous trees prevailing over the calcareous rocks, clothing especially the summits and N.E. slopes of the hills. Oak and beech are generally the most abundant, then follow the hornbeam, aspen, lime, sycamore, service-tree, ash and birch; the last occurs but sparingly: these deciduous woods are intersected by more or less considerable tracts of pine forest. The underwood is composed of a great variety of plants, among the most abundant of which are the hazel, maple, hawthorn, guelder rose, the wayfaring tree, hornbeam, dogwood, and goat willow. Most of the hedges and thickets which deck the slopes of the hills consist of these, and they harbour in their shade a host of the rarer plants. Among the shrubs, the lilac and honeysuckle are the most remarkable, giving a gay aspect to the bare declivities of the Saal in spring. *Mespilus germanica* is also common in this valley, and its vine-hills have quite a southern aspect when these, with the almond, peach and quince, are all in blossom together. The walnut flourishes best in the narrow side-valleys sheltered from the east wind, and there yields a valuable product.

The physiognomy of the vegetation on the sandstone formation is chiefly determined by the pine forests; and the plateau between the Saal, the Elster and the Roda is particularly distinguished by its extensive black woods of fir-trees. The Scotch fir predominates, the spruce sometimes scattered among it, or forming more or less extensive masses by itself. The silver fir occurs always

Q 2

solitary, as scattered through the Thuringian plains generally. The spruce fir is the prevailing tree in the
Thuringian forests.

The heather, which occurs in no very great extent,
and only associated with the fir woods on the sandy soil,
or on siliceous soils of the limestone district, has little
influence on the general character of the scenery. The
trees of the valleys, especially of the Saal, consist of
alders, willows and black poplars, which occur chiefly in
the meadows forming the bottoms of the main valleys ;
above them, on the lower part of the slopes, are found
the arable fields and plantations of fruit-trees ; and on
the upper portions the vineyards ; but the vine is only
cultivated on the Saal and in the side valleys, and does
not yield a very remarkable product—not nearly so
good as in some places further north, as around Naumburg.

It has been said that the climate of the valley of the
Saal is much milder than that of the surrounding provinces, and the mean temperature of the summer differs
still more, being higher than in any other part of Thuringia. The influence of this is seen in the prevalence
of the plants of the South German flora, and the small
number of northern species which occur there, while the
other parts of the district, particularly the elevated plains
and northern slopes of the hills, are characterized by the
opposite condition. The southern vegetation attains its
limits here, at the entrance of the Saal into the flat or
slightly hilly country below Naumburg, and is tolerably
well defined, both as to elevation and northern latitude,
by the limit of the vine (800–1000 feet). The watershed between the Saal and the Ilm also forms a tolerably
distinct line of demarcation between the plants of North

and South Germany. The following are some of the species which indicate the southern character of the valley of the Saal:—*Isatis tinctoria, Erysimum odoratum, Lepidium Draba, Dictamnus Fraxinella, Coronilla montana, Orlaya grandifolia, Stellera passerina, Aster Amellus, Centaurea solstitialis, Calendula arvensis, Melittis Melissophyllum, Anthericum Liliago, Andropogon Ischæmum,* &c. These and many others find their most northern limit below Naumburg, and the majority also disappear on the watershed between the Saal and Ilm.

The following species represent the North German flora:—*Viola palustris, Vaccinium Vitis-idæa, Oxycoccus* and *uliginosus, Ledum palustre, Eriophorum vaginatum, Schœnus nigricans,* &c. Thus it is mostly heath or bog plants that belong to the northern type, and they occur seldom elsewhere than in the S.E. parts of the Jena province, and are generally confined to the cold, wet moors in the district of the fir and pine woods.

The Hartz mountains present some curious peculiarities in their vegetation which deserve especial notice. In order to comprehend them in their true independence, we must exclude the low promontories and ranges of hills around the great mass, since the vegetation of these, both geologically and botanically, agrees with those of the terraces of the Elbe and Weser. And the calcareous plants of the northern borders of the Hartz correspond with those on the same rocks about Halberstadt and in Thuringia; but the most peculiar rock of this system, the richly clothed gypsum of the Zechstein formation, surrounding the south margins of the Hartz, occurs again, with the same plants, at Kyfhäuser and on the Bottendorf Height of the Unstruth.

Thus limited and separated from the stratified rocks

around, the Hartz forms a hilly plateau, connected even in the deeper valleys, of clay-slate and grauwacke, with isolated masses of igneous rocks. The uniformity of the structure, the rarity of lime on the surface, the small difference of elevation between mountain and plain, together cause a comparative poverty of peculiar forms of flowering plants. But on the other hand, a subalpine and even an alpine character may be detected, and this at a height above the sea, which would not lead us to expect it, and does not exhibit it nearly so clearly in the Saxon, Silesian and Rhenish mountains. That there is a peculiarity of the Hartz in this respect is most distinctly shown by the low position of the climatic tree-limit, above which the Brocken most probably rises a few hundred feet.

Similar phænomena are to be found on the western fjelds of Norway, where the conditions of climate are more readily comprehensible than in the Hartz. The spruce fir (*P. Abies*) grows up to a height of 5500 feet in the north of Switzerland (48° N. L.); in very favourable stations, here and there above 7000 feet. Its limit in Southern Norway, at Gousta (60° N. L.), lies at a level of 2900 feet. If the limit of the spruce ran down uniformly with the increasing latitude from Switzerland to Norway, it should stand at 4500 feet in the latitude of the Hartz (52°), and it does in reality attain 4400 feet in the fir zone of the Riesengebirge. Now in the Hartz it does not grow above 3300 feet, and thus suffers a depression of 1200 feet. The beech, again, will scarcely thrive above 2000 feet, yet in Northern Switzerland grows about to 4250 feet; so that this also has its level reduced by about 1200 feet.

Folgefond, in Bergen, lies in about the same latitude

as Gousta, but near the sea and surrounded by fiords. Here the tree-limits fall short of those of Gousta, which lies in the interior of the mountain region, as much as those of the Hartz do of those of the Riesengebirge ; all wood ceasing at 1800 feet, thus 1100 feet lower than at Gousta. These two mountains have therefore exactly the same relation to each other as the Hartz and the Riesengebirge, but they lie nearer together, and thus allow their climatal conditions to be more easily compared. The influence of the sea in decreasing the summer temperature is perceived and distinctly shown to be the cause of the depression of the tree-limits at Folgefond. The Hartz, in like manner, lies freely opposite to the sea, the north-western, prevailing currents of air from which strike upon it after passing over about thirty-five geographical miles ; and this is doubtless the cause of the depression of the tree-limits on these isolated mountains.

The following is a list of the Hartz species which do not occur, or but sporadically, in the surrounding districts. In the calcareous or granitic soil of the valleys : *Silene Armeria, Potentilla rupestris, Polemonium cæruleum, Aster alpinus,* &c. In the region above the tree-limit : *Anemone alpina, Aconitum neomontanum* and *variegatum, Alsine verna, Saxifraga cæspitosa, Myrrhis odorata, Sonchus alpinus, Armeria humilis, Salix bicolor, Betula nana, Carex rigida* and *vaginata,* &c. &c.

Ratisbon, situated in the plain of Bavaria, lies pretty nearly in the same annual isotherm as Jena, and the mean temperatures of the months do not differ much. The district around it, forming part of the basin of the Danube, presents a varied geological formation, consisting partly of primary rocks, principally granitic, continuous

with those of the Fichtelgebirge in the north, terminating on the south on the left bank of the Danube, and
stretching out to the east into the Austrian States; and
of secondary formations, composed of sandstones, liassic,
Jurassic and greensand rocks, a fragment of the formation extending over the western part of the Upper Pala-
tinate.   The sides of the valley of the Danube and of
the lateral valleys opening into it are mostly steep, the
south- and north-western declivities generally more precipitous than the south- and north-eastern.

Extensive pine-forests clothe a portion of the left bank
of the valley; the ridges and N.E. slopes are also extensively covered with the same, in unconnected masses;
while on the right bank cultivation has only left isolated and mostly small patches of pines.   The prevailing
tree is the Scotch fir; the spruce is much more rare,
occurring only in low, damp situations; the silver fir
and deciduous trees are usually sparingly scattered.   Of
the latter, the beech, birch, lime, aspen, sessile-flowered
oak and mountain ash are found here and there as inhabitants of the woods; the pedunculated oak, hornbeam, elm, sycamore, ash and service-tree are local and
rare.   Small woods of birch exist in many dry places
about villages, and the alder is planted in the swamps.
The larch and the mulberry have been successfully introduced within a recent period.   Only shrubby growths
of the above trees are met with on the steep north- and
south-western declivities, which are in most places so
much inclined, as to render them unfit for cultivation
either as fields or gardens; from Ratisbon downwards
there exist on several of them vineyards of great antiquity, which still repay the care bestowed on them; but
this branch of cultivation has not been developed to the

extent that seems possible by selecting varieties of the vine properly suited to the climate.

The character of the vegetation will be understood by an enumeration of the commonest plants growing in the various stations which the district affords. On the bare cliffs there are *Arabis arenosa, Alyssum montanum, Sedum album, Allium fallax* and *flexum, Holcus australis, Stipa pennata* and *Sesleria cærulea,* all plants of a southern type. The dry slopes of the hills when unaltered by cultivation, particularly those immediately forming the banks of the Danube and its tributary streams, afford the most characteristic plants of the Ratisbon flora. The prevailing species here are : *Clematis recta, Anemone Pulsatilla, Erysimum odoratum* and *crepidifolium, Isatis tinctoria, Viola collina, Polygala Chamæbuxus, Silene nutans* and *Otites, Geranium sanguineum, Cytisus nigricans* and *ratisbonensis, Cerasus Malaheb, Cotoneaster vulgaris, Seseli coloratum, Libanotis montana, Peucedanum Cervaria, Chrysocoma Linosyris, Aster Amellus, Buphthalmum salicifolium, Centaurea paniculata, Lactuca perennis, Hieracium Nestleri, Euphrasia lutea, Stachys recta, Prunella grandiflora, Thesium montanum, Euphorbia verrucosa, Orchis militaris, Anthericum ramosum,* and *Andropogon Ischæmum*; a list which might serve to characterise the warm and sheltered parts of the valleys of the Moselle and Middle Rhine in the west of Germany.

At the foot of the hills, and here and there upon their slopes, are found hedges and bushes in which the common clematis, berberry, maple, buckthorn, blackthorn, hawthorn, wild cherry, elder, privet, hazel, juniper, &c. abound, while the bladder senna, *Cratægus monogyna,* and the black poplar are rare. The commonest plants inhabiting the shades of these are *Sedum maximum, Symphytum tuberosum, Cerinthe minor, Verbascum Blattaria, Gagea lutea,* and *Scilla bifolia.*

The woods, principally composed of the Scotch fir, possess a very rich flora. The other trees before-mentioned are sparingly sprinkled among these with most of the common shrubs of Northern Europe, together with the heather, the whortleberry, and the southern heath, *Erica carnea.* The Compositæ, Leguminosæ, Orchidaceæ, Grasses and Rosaceæ abound most here. Among the most characteristic of the common plants are :—*Anemone Hepatica, Ranunculus nemorosus, Aconitum Lycoctonum, Cytisus capitatus, Rubus saxatilis, Astrantia major, Peucedanum Chabræi, Erica carnea, Pyrola secunda, Lithospermum purpureo-cæruleum, Digitalis grandiflora, Melampyrum nemorosum, Melittis Melissophyllum, Mercurialis ovata, Cypripedium Calceolus, Lilium Martagon,* and *Carex longifolia. Dianthus Seguierii, Lathyrus heterophyllus* and *Mercurialis ovata* are found in few other floras besides that of Ratisbon. Among these woods, and in various other spots, occur marshy meadows, chiefly overgrown with sedges, which form one-third of the species inhabiting them. In some places they are decked with alder bushes, and the most remarkable plants are *Trollius europæus, Polygala amara, Dianthus superbus, Phyteuma orbiculare, Gentiana verna, Pinguicula vulgaris, Primula farinosa,* and *Triglochin palustre.* The following are found only in the extensive marshy meadows in the western part of the district : *Ranunculus Lingua, Viola palustris* and *stagnina, Stellaria glauca, Lathyrus palustris, Senecio paludosus, Mentha Pulegium,* and *Schœnus ferrugineus.*

Where the woods have been thinned or removed, the

fertile soil has been converted into pasture or arable land; and as the sedges prevail in the marshy spots, so on the drier meadows and rich pastures the grasses acquire the predominance both in number of species and of individuals; forming about one-fifth of the flora. After the grasses, the commonest plants are: *Thlaspi perfoliatum*, *Tunica Saxifraga*, *Linum perenne*, *Cytisus sagittalis*, *Ononis spinosa*, *Coronilla varia*, and *Campanula patula*.

The common crops in the cultivated fields are: rye, wheat, two-rowed barley and oats, potatoes; less frequent are turnips, cabbages, coles, &c.; and more rare still, lucerne, saintfoin, beet, flax and hemp; while the six-rowed common barley, millet, maize, buckwheat and tobacco are seldom sown. Among the commonest weeds are: *Sisymbrium Thalianum*, *Nestlera paniculata*, *Silene noctiflora*, *Lathyrus tuberosus*, *Falcaria Rivini*, *Caucalis daucoides*, *Anthemis austriaca*, *Prismatocarpus Speculum*, *Myosotis striata*, *Veronica didyma*, *Amaranthus retroflexus*, *Passerina annua*, *Gagea stenopetala*, *Muscari comosum*, &c.

About the banks of the rivers the alders and willows form more or less dense thickets, among which *Stenactis annua*, *Veronica longifolia* and *Rumex maritimus* are especially abundant. *Cucubalus bacciferus*, *Aster Salignus*, *Senecio sarracenicus*, *Carduus Personata*, *Rumex maritimus*, *Hippophaë rhamnoides*, *Alnus incana*, *Scirpus trigonus* and *triqueter* are only found in the Danube itself.

Places liable to be periodically overflowed are not very common in the district; where they do occur, and especially on the damp sand of the river's banks, *Erucastrum Pollichii*, *Œnothera biennis* and *Juncus bufonius* are the commonest species. True water-plants are very numerous, amounting to 60 species.

The flora of Ratisbon contains in all 1063 species of flowering plants, belonging to 108 families; the German families not represented in it are the Capparidaceæ, Zygophyllaceæ, Terebinthaceæ, Cæsalpineæ, Granateæ, Tamariscineæ, Myrtaceæ, Cactaceæ, Lobeliaceæ, Ebenaceæ, Aquifoliaceæ, Polemoniaceæ, Acanthaceæ, Lauraceæ, Cytineæ, Myriceæ, Dioscoreæ and Bromeliaceæ, which mostly belong exclusively to the extreme south of Germany, and of which only a few South-European species extend over into the flora of the north side of the Alps.

The principal families are : the Compositæ $= \frac{1}{9}$, Grasses $\frac{1}{13}$, Sedges $\frac{1}{16}$, Leguminosæ $\frac{1}{18}$, Labiatæ and Cruciferæ $\frac{1}{22}$, Umbelliferæ $\frac{1}{23}$, Caryophyllaceæ $\frac{1}{25}$, Scrophulariaceæ $\frac{1}{26}$, Ranunculaceæ $\frac{1}{28}$, Rosaceæ $\frac{1}{35}$, Orchidaceæ $\frac{1}{48}$, Boragineæ $\frac{1}{53}$, Polygonaceæ and Liliaceæ $\frac{1}{55}$, Salicaceæ $\frac{1}{58}$, &c. &c.

The Carpathians.

Our knowledge of the vegetation of the Carpathian mountains extends only to the northern part of the range, in the north of Hungary; and it is here that the highest points, the Lomnitz and Eisthal summits, are situated. Space will only allow of a brief and general account of this region, for comparison with the other important elevated regions of Europe.

The vine can nowhere be cultivated above about 900 feet in the lands surrounding the Carpathians, and the walnut only ascends to 1232 feet; so that the region may be considered colder in respect to vegetation, although not much more elevated, than Northern Switzerland. The plants of the vine and walnut region are thus excluded from the Carpathian flora; yet this exhibits as many and as beautiful plants as the flora of the north of Switzerland, since the higher regions, possessing a more continental climate than on the Alps, produce a more abundant and attractive vegetation.

The cultivation of grain and orchard fruits extends higher than in Switzerland, barley and rye covering large tracts upon the lower part of the slopes, not however penetrating much into the internal valleys.

1. The *mountain region*, characterized by the presence of the beech, extends up to about 3900 feet, apparently about the same altitude as it attains on Mont Rigi and Mont Pilatus. At the upper border of this region much the same alpine plants occur as in Switzerland, and these do not descend much lower down in Hungary than around the Alps. This region exhibits a very rich and luxuriant vegetation. Although the evergreen shrubs, such as holly, ivy, *Polygala Chamæbuxus*, &c., are not found, their loss is more than compensated by the abundance of herbaceous plants of large size, such as *Senecio umbrosum, Symphytum tuberosum, Astrantia Epipactidis*, &c., and the brilliantly coloured *Geranium phæum, Cortusa Matthioli, Lilium Martagon*, &c. Where the soil is free from wood, wonderfully rich meadows and pastures are seen, in which the perennial grasses, *Poa sudetica* and *Avena planiculmis*, are conspicuous. Many herbs common to both countries extend higher up than in Switzerland, while such ligneous plants as the maple, *Viburnum Lantana*, and others are arrested lower down.

2. The *subalpine region*, between the limits of the beech and the spruce, exhibits in its lower part a vegeta-

tion similar to the corresponding region in Switzerland; thus the sycamore, *Sambucus racemosus*, &c., and here and there *Veronica aphylla, Ranunculus alpestris, Carex firma*, &c. are met with. But the character of the region is soon totally changed, the dull and useless *Pinus Mughus* covering the whole surface in place of the rich pastures of subalpine Switzerland; the spruce is soon after lost, the specimens of some six feet high marking the limit at 4600 feet on the outer slopes of the range, and not rising higher than 4700 feet in the cross valleys of the centre. In Switzerland the spruce ascends to 5500 feet, 900 feet higher than in the Carpathians; so that while this subalpine region has a perpendicular range of 1400 feet in the former, it is narrowed to 665 feet in the latter; nor can the vegetation at the upper limits be said to have a more delicate character in the Carpathians, since *Empetrum nigrum, Pedicularis versicolor,* and a little above, *Anemone alpina*, occur as examples.

3. The *inferior alpine region*, from the limit of the spruce fir to that of the *Pinus Mughus* growing some two feet high in open places, is a very naturally defined region in the Carpathians—much more so than in Switzerland or Lapland; for though shrubs of the green alder afford us a mark in the former and the glaucous willow in the latter, these occur much more scattered than the *Mughus*, which is here almost universal. Beneath the shade of the *Mughus* grow many large and handsome plants, such as *Doronicum austriacum, Cortusa Matthioli, Cineraria crispa, Hypochæris helvetica, Swertia perennis,* and many field plants, which occur higher here, under the shelter of the *Mughus*, than in Switzerland; exposed declivities however, devoid of this shrub, afford scarcely any herbage, almost the only socially growing

plant being *Poa disticha* ; but such species as *Campanula alpina, Senecio abrotanifolia,* &c., mark the character of the region. Around the alpine lakes, especially where steep overhanging banks keep in the watery vapours rising under the sun's influence, a wonderfully rich vegetation is found in summer. The upper limit of the *Mughus,* in a growth of about two feet high, occurs at about 5600 feet, and up to this point it forms almost everywhere a complete covering to the surface. Higher than this a few isolated plants do indeed occur, as far as 6000 feet, but they are starved or creeping among the rocks.

4. The *upper alpine region* loses, with the *Mughus,* almost all the beautiful plants that grow beneath its shelter, and on the mountains of Tatra, in the west part of this range, it is so dry and sterile as to remind the beholder of the upper regions of Lapland ; but in the latter, the subalpine tract which has here so barren an aspect is covered with abundance of small heath-like shrubs, such as *Andromedas, Azaleas, Diapensia,* &c., forming a complete carpet, while in the Carpathians scarcely any occur, and the surface of the subalpine tract is bare and stony, completely sterile save where a little accumulation of earth nourishes a tuft of *Poa disticha,* or *Senecio abrotanifolia* with its leaves become almost spiny ; or here and there are scattered solitary plants of diversified habit, but with large and brilliant flowers, such as *Arnica Doronicum, Primula minima, Campanula alpina, Gentiana frigida, Dianthus alpinus,* and the strangely succulent *Serratula pygmæa,* which makes one wonder how its soft and fleshy leaves find nourishment on the dry rock. Most of these plants are of succulent habit, differing in this respect from the plants of other mountains ; their roots are particularly large and fleshy. The long distance between the

limit of the *Mughus* and the snowy region, measuring 2400 feet, is covered with this vegetation on its lower half; above, *Empetrum, Vaccinium uliginosum* and *Salix retusa* are intermingled. But the upper portion, above 6500 feet, and especially on the summits of Tatra, is barren and desolate. These peaks are described by Wahlenberg as the poorest in plants of any alps he had visited; the rocky surface is broken and covered with loose blocks, and though these are of whitish granite, black lichens overgrow them all, so that, devoid either of snow or herbage, the peaks stand up like pitchy cones, on which long streaks of white appear, caused by the washing away of the lichens by violent rain-storms. On the peak of Krivan only ten flowering plants were found, and this has the mildest climate of any of these peaks; on the summit of the Lomnitz, Wahlenberg found only *Gentiana frigida, Saxifraga bryoides, Ranunculus glacialis,* and *Poa disticha,* and these in such small quantity that he carried away the whole of the vegetation of the extreme summit in one hand; a striking contrast with the conditions met with on the Alps.

5. The *snowy region.*—This can hardly be said to be truly developed on the Carpathians, since the snow almost always melts off nearly all the peaks in summer, and the only traces that remain are accumulations in hollows and ravines, or in the small valleys of the higher masses; small glaciers are thus produced, and these furnished the principal grounds on which Wahlenberg established his snow-limit at 8000 feet on the Eisthal peak. It would seem more accurate, however, to reject this altogether from the Carpathians.

There are very considerable local differences within the Carpathian range, both in the nature of the species and their range of altitude, but we cannot here enter into details on this head. It may be observed, however, that many plants of the subalpine region occur on the outer parts of the chain which are wanting on the central mountains, as *Gentiana acaulis* and *verna, Dryas octopetala, Androsace pauciflora, Arenaria laricifolia*, &c. ; while in the central masses this region contains two trees, the larch and *Pinus Cembra,* not elsewhere met on the chain. With regard to the character of the floras on the different sides of the chain, the lower grounds on the east exhibit a vegetation more approaching to that of the Hungarian plain than any other part, as is shown by the presence of *Artemisia Scoparia, Seseli annuum, Rosa pumila, Nepeta nuda, Sisymbrium pannonicum, Vicia villosa* and *pannonica,* &c. The accessory mountains on this side also afford abundance of many peculiar species, such as *Astragalus alpinus, Phaca alpina, Hedysarum alpinum, Ranunculus Thora, Gentiana glacialis* and *nivalis, Draba tomentosa* and *pyrenaica, Cineraria capitata, Saxifraga oppositifolia, Serratula alpina, Androsace villosa,* &c., not found elsewhere in the Carpathians ; while still more locally, in one of their valleys, have been found *Primula longiflora, Phaca australis, Cerinthe maculata, Euphrasia salisburgensis* and others. The low mountains on the south have *Arbutus Uva-ursi, Erica herbacea, Daphne Cneorum, Campanula carpatica, Cytisus ciliatus,* &c. Those on the west exhibit more of the Austrian and Swiss character than any other part of the region ; thus among their characteristic plants are : *Saxifraga rotundifolia, Dentaria enneaphylla, Cardamine trifoliata, Viola alpina, Fumaria capnoides, Veronica montana, Buphthalmum salicifolium, Aconitum Lycoctonum,* and on the east side of Tatra, *Tozzia alpina.*

R

The general differences of the Carpathian vegetation and that of the north of Switzerland seem to be, that there is greater local diversity in the Carpathians; the richer vegetation of its two flanks disappears much more quickly toward the central masses, that is, the altitudinal limits are lower, so that the summits are exceedingly poor and barren, having a Piedmontese or Tyrolese character rather than a Swiss. The spruce forests terminate much lower down, so that the alpine region is comparatively more extensive, having a range of 3400 feet, while in Switzerland it has only 2708, and in Lapland between the birch and the snow only 1500 feet.

The principal reasons for these peculiarities appear to lie in the situation with regard to the isothermal lines, and the distance from the ocean, producing the excessive or continental character of climate; these, and the absence of an accumulation of eternal snow, dependent on the deficient altitude under such a summer isotherm, depriving the upper region of that continuous supply of moisture in which the Alps abound, produce an effect in some degree analogous to that which is met with on the barren summit of Etna.

Sect. 7.

The Alps.

THESE, the highest mountains of Europe, stretch in an uninterrupted mass from the Hungarian plain to the mouth of the Rhine; passing into the Dinaric mountains at the south-east end, and the Apennines at the south-west. They are bounded naturally on the south by the great plain of Lombardy (the valley of the Po), on the west by the lower part of the course of the Rhine, on the north-west by a depression separating them from the Jura, and on the north by the valley of the Danube. Only a very small portion touches the sea-coast.

The Alps lie between the $43\frac{1}{2}°$ and $48°$ of N. L., and are 150 geographical miles in length; the breadth varies from 20 to 40, and decreases towards the west. The principal direction is from E.N.E. to W.S.W., but the western portion runs N.N.E. and S.S.W.

The following data give some idea of the mean elevation:—

	Feet.
From the S.W. extremity to Mont Viso...	6,400
From Mont Viso to Mont Blanc	9,000
From Mont Blanc to Monte Rosa........	11,800
From Monte Rosa to the Brenner........	9,600
From the Brenner to the Great Glockner..	7,000
From the Glockner to the N.E. extremity .	4,700

As a whole the western portion is loftier than the eastern, and the highest point lies nearer to the S.W. end than to the N.E. limit.

Among the highest peaks are:—

	English feet.
Monte Viso	12,547
Loucyra	14,394
Mont Blanc	15,722
Monte Rosa	15,165
Jungfrau...............	13,663
Finsteraarhorn	14,061
Ortler.................	12,812
Great Glockner	13,263
Terglou	9,874
The Steiner Alp.........	10,916

The passes are numerous: the Stilfser Joch, 9148 feet, and the Great St. Bernard, 8148 feet above the sea, are the highest; the lowest are in the Tyrolese Alps, the Brenner being 4886 feet, and the Semmering 3317.

The highest parts of the Alps are towards the south side of the mass, which, on the whole, is more steep than the north; the great Bavarian plateau forms a sort of terrace 1700 feet above the sea on the north side. Large valleys run lengthways among them, such as the Valais, the upper valley of the Inn, and the Veltaline; but the cross-valleys are more numerous, such as the Val d'Ossola, the Levantine and the Etschthal on the south side, and the valleys of Hasli, Reuss, Iller, Lech, Salza, &c. on the north. Large valleys run out at the east extremity, from whence flow the rivers Mus, Drau, Gail and Sau; the Tinca valley runs out from the S.W. extremity.

The most important rivers rising in the Alps are the Po, originating in the Maritime or S.Western Alps, which receives nearly all the waters of the southern slopes, and empties itself into the Adriatic; the Rhine, springing from the north side of the Alps and flowing northward between the mountains of Central Europe; the Rhone, also rising on the north side, but flowing round a part of the range to the Mediterranean; and the Danube, the original sources of which are indeed in the Black Forest and Albe, but which is principally fed by the waters of the north and east sides of the Alps.

Considerable lakes exist at the foot of the Alps: on the south side the Lago Maggiore, Lago di Como, Lago di Guarda, and many others; on the north, the Lake of Geneva, of Neufchatel, of Zurich, and Constance, 1200–1400 feet above the sea, and others; on the east, the Neusiedler and Platten Lakes; while none of importance occur on the west. Smaller lakes occur at greater elevations, such as that upon Mont Cenis, 6100 feet above the sea level.

The nucleus of the Alps is formed of old rocks, such as granite, mica-slate, granular limestone, &c., and the highest peaks of one of these formations; on each side of these are applied more recent rocks, especially mountain limestone, of which a very considerable part of the Alps is composed. Volcanic masses are rare.

The following table gives a summary of the temperature :—

	Lat.	Elev.	Annual.	Winter.	Summer.
	°	ft.	°	°	°
Avignon . . .	44		39	$42\frac{1}{2}$	74
Marseilles . .	$43\frac{1}{2}$		58	47	68
Milan	$45\frac{1}{2}$		55	36	73
Ofen	$47\frac{1}{2}$	506	51	31	70
Geneva	46	1275	50	35	63
Munich. . . .	48	1730	48	$30\frac{1}{2}$	65
Peisenberg .	48	3281	42	29	58
St. Gothard	48	6841	30	17	$43\frac{1}{2}$
St. Bernard	$46\frac{1}{2}$	8148	31	$18\frac{1}{2}$	$43\frac{1}{2}$

Thus the annual mean temperature of the S.W. extremity of the Alps, where they sink towards the Mediterranean (Avignon, Marseilles), is considerable, and the winter is particularly mild there. It is lower in the plain of Lombardy (Milan), the winters being comparatively cold there, in fact, colder than in Edinburgh, while the summer heat is high. In a series of observations carried on through many years, the highest temperature of Milan was $92\frac{3}{4}°$, and the lowest $5°$. The maritime Alps and the Apennines intercept the influence of the sea, and give the plain of Lombardy a climate approaching the continental character. The seasons are still more diverse in the eastern portions of the Alps, on account of the inland position; the difference of summer and winter in Marseilles is only 21°, in Milan it is nearly 37°, and in Ofen 39°. The Bavarian plateau, lying 1700 feet above the sea, has the climatal condition of Denmark, and the thermometer sometimes falls to $-19\frac{3}{4}°$, while in Copenhagen it has not reached $-9\frac{3}{4}°$ during fifty years. The mean temperature of Geneva, also on account of its elevated position, is lower than that of Paris, although the latter lies 3 degrees to the north; but the winter is tolerably mild in Geneva, because the valley of the Rhone lies open to the warm S.W. winds. At an elevation of 3000 feet (Peisenberg) on the Alps is found the mean temperature of Stockholm, but a milder winter and a cooler summer than in that city. At St. Gothard and St. Bernard, from 6000–8000 feet, the mean temperature is lower than at the North Cape. Thus in the Alps we may pass in a few days through as many different climates as lie between these and the northern limit of Europe. It is estimated that the mean temperature of Mont Blanc, the highest summit in Europe, is 27° below the freezing point.

A large quantity of rain falls at the southern foot of the Alps; a mean of numerous observations at different stations gives 50 to 60 inches, and at some points in Friuli it amounts to nearly 100 inches. The southern and S.W. winds bring a great deal of vapour, coming from the sea and from warmer regions where evaporation goes on more actively; when these currents, loaded with aqueous vapour, strike against the cold Alps, the water is precipitated in frequent and considerable rains. The fall is not so great on the west side, but increases with the advance into the interior of the range. The same holds good, on a smaller scale, of the north side. The smallest quantity of rain falls on the east side of the Alps; Ofen, for example, has only 18 inches: this is accounted for by the inland position, and the fact that the southern and south-western winds come from the Alps over the warmer Hungarian plain.

The snow-line lies at 8200 feet above the sea on the north side of the Alps; on the south side it rises to 9500 on Mont Rosa, but sinks to 8000 on the east; the mean of the south side may be estimated at 8600 feet. Thus it is seen that a very considerable portion of the Alps is covered by eternal snow; and the snow-covered peaks of this vast range are seen in this condition at the hottest season of the year from the plains of Lombardy and Bavaria. From these great fields of snow pass glaciers into the ravines and valleys, enormous masses of ice sinking gradually downwards and melting there, their lower extremities marked by a barrier formed of stones and debris, from which flow streams of milky water, as in the Scandinavian and Icelandic mountains. The most numerous and extensive glaciers occur in Savoy, especially in the Valley of Chamouni below Mont Blanc, in the Tyrol, in the Bernese Oberland, and in the Grisons. They come down to within 3000 feet above the sea, and thus sometimes lie on a level with corn-fields.

The character of the southern foot of the Alps will be described in speaking of Italy, but it is necessary to notice especially here the region at the western extremity, the Mediterranean district of France. The vegetation of this region belongs completely to the Mediterranean type, being analogous to that of Spain and Italy, and quite distinct from that of the centre and north of France.

Here, and along the coast at the foot of the Maritime Alps, are found the orange, myrtle, cactus, dwarf palms, &c., as in Andalusia and Naples, and among the indigenous plants the Leguminosæ acquire that predominance which marks the southern climate. It is not necessary to dwell upon this at present; it will suffice to direct attention to the character of the declivities of the Alps in this region. On Mont Ventoux, for example, the Aleppo pine, with the olive, *Quercus coccifera, Genista hispanica*, &c., are found, rising to about 1400 feet. From 1400 to 1800 feet extends the evergreen oak, with *Centaurea solstitialis, Scolymus hispanica, Xanthium spinosum, Juniperus oxycedrus*, &c. From 1800 to 3800 feet there are no trees; thyme and lavender cover the ground, mixed with dwarf box, *Nepeta graveolens, Aphyllanthus monspeliensis, Carlina acanthifolia*, &c. The beech region reaches from 3800 to 5500 feet; the trees at first dwarf and scattered, then larger and in groups. Under their shade grow northern species, mingled with a few subalpine.

Below this limit *Pinus uncinata* joins the beech, and from 5500 to 6000 feet it reigns alone; in this region most of the plants of the last occur, with the addition of *Teucrium montanum, Saxifraga cæspitosa* and *Juniperus communis*. Lastly, the alpine region extends from 6000 to 6330 feet, the summit; in this the plants are completely alpine, such as *Viola Cenisia, Oxytropis cyanea, Saxifraga oppositifolia, Poa alpina*, &c.; and what is interesting to observe, *Helianthemum œlandicum* is found here in the alpine region, while in the Swedish island of Bornholm it occurs little above the sea-level; a striking example of the relation between altitude and latitude.

The northern faces of the mountains in this district do not possess the Aleppo pine at their feet, which are always at least 1300 feet above the sea; but the evergreen oak, the chestnut, the lavenders and thymes, the beech, the *Pinus picea* and *uncinata* are successively passed in proceeding up to the alpine region.

In regard to the other parts of the Alps, where the local circumstances do not produce such marked peculiarities, we must be content to examine the different regions in a very general manner, taking a kind of average character, and mentioning special singularities only when of great importance.

The following descriptions, therefore, apply to the Alps generally, excepting the maritime western extremity.

The region of the plains or lower hills is limited above at about 1700 feet, where it is succeeded by the *lower mountain region* or zone of the chestnuts, extending up to 2500 feet. Throughout the Northern Alps this is tolerably even in the different chains. In the Central Alps the lower zones of vegetation can only be observed in the deep main valleys, as the bottoms of the other valleys lie at a great height. The chestnut however reaches a great elevation in some of the high basin-like valleys which are fully exposed to the sunlight, especially on their southern slopes; the maximum elevation noted is 3600 feet. Important modifications present themselves in the Southern Alps, since the climate of Italy exerts great influence here, and a distinct series of zones of vegetation appear with the olive, stone pine, fig, laurel, &c.; even the vine and chestnut ripen higher on the south face of Monte Rosa than the chestnut generally does in the Northern Alps.

In this region cultivation not only includes all general European grains, but also maize, and up to a certain point the vine. The character of its vegetation therefore, as might be expected, agrees with that of the Central European region already described on the north side, while on the south there is an infusion of southern species from the Italian flora. No minute details therefore will be necessary here.

The *upper mountain region* or zone of the beech extends on the south side from 2500 to 4000 feet, and from 2000 to 4000 on the north. This may be also regarded as the zone of the higher deciduous trees; it is far more developed in the Northern than in the Central Alps, the upper limits also exhibiting great irregularity in the latter. Perhaps the ash and the cherry are the most evenly distributed trees, affording tolerably exact limits as accompanying the last villages; those of the beech, oak and maple being exceedingly variable they are rather rare, and never form large woods in the Central Alps, and are often absent in the inner parts of the higher groups even at a low elevation. The limit of the beech especially, taking groups of trees as the test, is not generally so high on the south as on the north side of the Alps, and undergoes remarkable depression in many places. Those of the oak, the maple and the bird-cherry have about the same average height. In the middle and upper limits of the Cerealia the inclination of the mountains is of great importance, since in addition to the elevation of temperature and the greater summer heat, the gently inclined bottoms and sides of the valleys, capable of cultivation up to a great height, greatly favour the culture of grain, in contrast to the more precipitous valleys and declivities of the lower chains at equal heights. The effects of the southern and south-western exposure are strongly marked in the upper limits of the Cerealia.

Different grains occupy the uppermost place in different places; this depends partly on the wants of the inhabitants and partly on custom, but some distinctions are also caused by the distribution of temperature and moisture. In general barley and oats are grown the highest, but winter rye sometimes occurs as high; wheat always ceases lower down.

The indigenous vegetation here again presents the analogy to the North-German flora which the preceding does to that of Central Germany; at the same time local peculiarities give rise to many subordinate differences; the subalpine plants of the preceding region here and there descend into this and modify its character; the absence of suitable localities excludes many of the aquatic or marsh plants of northern Europe, and among other points may be noted the entire absence of the bog myrtle (*Myrica Gale*) and the Lobelias of northern Europe.

The *subalpine region* or zone of the Conifers extends from 4000 to 5500 feet on the north, and 4500 to 6500 on the south side in general; but great variations exist, depending on inclination, exposure to storms, &c. The Scotch fir is least widely distributed and most variable in its limits; the limits of the spruce fir, the larch and Siberian pine also exhibit variations, differences of several hundred feet being often produced even by the vicinity of glaciers; this however occurs chiefly near isolated glaciers, and is not so evident in the average limits in the Central Alps.

Human habitations are sometimes situated as high as the upper limit of the Coniferæ, in the Central Alps rarely as high as 6000 feet; in the southern parts little villages sometimes lie between 6200 and 6300 feet; potatoes, cabbages and turnips are grown with difficulty at these stations.

The condition of the vegetable covering of the soil in this region is very much dependent on the physical condition of its surface. The alpine pastures, so bright and green in summer, decked with so many showy flowers, are confined to the more gentle slopes and the sides of the mountain valleys, where the products of the gradual disintegration of the rocky masses, and the débris of innumerable generations of plants have accumulated and clothed the hard surface with a covering of mould. The fir woods also vary much in their distribution, especially where they have been interfered with by man's agency. In many parts of Switzerland the forests have been in great part destroyed, and hence a barren character is given to the localities. In the Tyrol the fir woods are very beautiful, climbing as it were in long lines up the sides of the steep mountains, finding foothold on the narrowest shelves, or where the declivities are less abrupt spreading out in black masses which give a remarkable grandeur and solemnity to the scenery.

> But from their nature will the Tannen grow
> Loftiest on loftiest and least shelter'd rocks,
> Rooted in barrenness, where nought below
> Of soil supports them 'gainst the alpine shocks
> Of eddying storms; yet springs the trunk and mocks
> The howling tempest, till its height and frame
> Are worthy of the mountain from whose blocks
> Of bleak, grey granite into life it came,
> And grew a giant tree.

The firs diminish in height as the elevation increases, finally dwindling down to mere shrubs of four or six feet in height; the grass grows more sparingly, the stony surface of the rock becomes more and more widely exposed, and the transition is gradual towards the next region, where trees are seen no more, and where the ligneous vegetation assumes a totally different character.

This region offers a striking contrast to the analogous zone in Scandinavia, where the fir woods are exceedingly poor in vegetation. Here the flora is wonderfully rich and varied; the botanist finds a rich harvest of the mountain plants with their large and brightly coloured flowers, rejoicing in the brilliant sunlight of these elevated tracts. The northern botanist recognizes numbers of his own subalpine forms, such as *Dryas octopetala, Saxifraga oppositifolia, Alchemilla alpina, Gentiana verna,* &c. &c.; but a large number are added, partly common to the higher mountains of North and Central Germany, partly common to the Pyrenees, and partly peculiar to the Swiss Alps.

A considerable difference is seen according as the mountain is of crystalline or calcareous structure, the former exhibiting a richer covering of vegetation, but the latter generally possessing a greater variety of species.

The *alpine region* or zone of shrubs succeeds, lying between 5500 and 7000 feet on the north side, and commencing about 6500 feet on the south face of the Alps. This region is very irregularly developed in the various groups, different members of its flora being predominant in different localities.

The grassy pastures are almost lost as we enter this zone, small patches of herbage occurring in more sheltered spots, to some of which the sheep and goats find their way in the summer season. But the general character of the surface is stony, and where it is not clothed by the often extensive patches of the low alpine shrubs, the herbaceous plants isolated and scattered, growing in tufts and patches, would scarcely remove the appearance of sterility were it not for the large size and brilliancy of their blossoms.

The alpine shrubs have a very peculiar aspect; the tendency to upright growth seems lost, and they creep along, winding and twisting their tough wiry branches, little above the surface of the ground, interlacing often into dense masses, and forming a coating somewhat similar in aspect to the heathy covering of the sandy plains of the lowlands. The most remarkable among these are the rhododendrons or alpine roses as they are called, which are the glory of this region. The species *R. hirsutum* is most common in the Northern Alps, especially on limestone, often clothing large tracts, intermingled with the strange-looking, gnarled dwarf pine, the 'Knie-holz' of the Germans. The *R. ferrugineum* is predominant in the crystalline slates of the Central Alps, mingled with juniper, alder, &c.

It will be remembered that a zone of dwarf birches occurs above the firs in Scandinavia; these are not found in this situation on the Alps,—their place seems to be taken by the rhododendrons; but the dwarf growth of the alder (*A. viridis*) may be regarded as their representative in the Amentaceous family.

The *subnival region*, 7800 to 8500 feet, comprehends the tract between the limits of the shrubby plants and the snow-line. A very large number of plants are still found here, which, lying buried beneath the snow for a long period of the year, break forth into a short life during the summer months when their frozen covering is temporarily removed. The low shrubby plants, the *Rho-*

dodendra, Vaccinia, Azalea, &c., ascend a little way into this zone, and the *Hieracia* do not reach much higher. Among the genera most abundantly represented here are *Saxifraga, Gentiana, Primula, Poa, Ranunculus,* and *Draba*. The Compositæ, Grasses, Leguminosæ, Cruciferæ, Caryophyllaceæ and Ranunculaceæ are among the most numerous families, while of the Umbelliferæ there are but few, such as *Meum Mutellinum, Gaya simplex,* &c.; of Rosaceæ only two or three Alchemillas; few Labiatæ, *Ajuga pyramidalis* being the usual representative of its family.

But the distribution of the plants at great elevations is subject to much modification from local conditions of the surface; thus in the vicinity of glaciers, or on very steep cliffs, the limits will be brought lower than on easy slopes; sheltered hollows exposed to the sun's rays, where earth has accumulated, are always especially rich in alpine plants.

On the Alps of Glarus as many as 219 species of flowering plants have been gathered in this region; in the Möll district, in the eastern Alps, Messrs. Schlagintweit collected 224 species between 7000 and 8500 feet; yet in remarking these considerable numbers we must recollect, that the plants exhibit still more than those of the preceding zone the effect of the alpine climate, in the scattered, isolated and independent manner of their growth; they are found nestling in the chinks of the rocks, or forming little turfy clumps on the accumulations of soil in the hollows, and from their small size they exert but a comparatively insignificant influence upon the scenery around, the features of which are marked with such grand outlines. It is very different from the influence of vegetation upon the scenery of plains, where the

earthy foundation is altogether hidden by the perfect
carpet of vegetation, and the giants of the vegetable kingdom, the forest trees, become by their size and number the
most striking objects of the landscape. This of course
applies chiefly to the fertile plains of Central Europe; in
the arid plains of the south, scorched by the summer's
sun, we find again a temporary condition somewhat analogous, in respect to vegetation, to that met with on the
Alps; on the one hand the result of intense heat, on the
other that of intense cold.

Within the snow-limits, above the line which marks
the average level of the lower edge of the 'eternal snows,'
at about 8500 feet, isolated, acute peaks or steep cliffs
rise here and there so suddenly from the general surface,
that the snow cannot lodge in sufficient quantity upon
them to withstand the influence of the summer heat, so
that for a few weeks they lose their frozen covering. On
such spots flowering plants have been observed up to a
height of as much as 11,350 feet on Monte Rosa, and
on several other summits above 10,000 feet. Messrs.
Schlagintweit found 33 species between 8500 and 10,000
feet in the Tauern of the Möll district, many of which
they have in common with the higher mountains of the
north of Europe; such as *Silene acaulis, Alsine verna,
Cherleria sedoides, Arenaria ciliata, Cerastium alpinum*
and *latifolium, Saxifraga oppositifolia, Azalea procumbens, Salix herbacea, Poa alpina*; some occurring in Iceland and Scandinavia, but not in Scotland, as *Ranunculus glacialis, Salix retusa*, and *Poa laxa*. Of the 33
species, 24, viz. *Ranunculus glacialis, Silene acaulis, Alsine verna, Arenaria ciliata, Cerastium alpinum, Sempervivum arachnoideum, Saxifraga androsacea, biflora,
bryoides, cæsia, muscoides* and *oppositifolia, Artemisia*

mutellina and *spicata, Chrysanthemum alpinum, Phyteuma pauciflorum, Azalea procumbens, Androsace glacialis (pennina,* Gaud.), *Salix herbacea* and *retusa, Carex firma, Avena subspicata, Poa alpina* and *laxa,* occur also in the Pyrenees. *Cerastium latifolium* occurs in the north of Europe, but apparently not in the Pyrenees. The remaining 8 are found neither in the Pyrenees or the Scotch and Scandinavian mountains : viz. *Cherleria sedoides, Mœhringia polygonoides, Gaya simplex, Gentiana bavarica, Pedicularis rostrata, Primula minima, Sesleria microcephala* and *sphærocephala.*

The species which occur at the highest altitude are not the same in all cases ; in the larger groups of the Alps they almost all reach to between 10,000 and 10,500 feet. In different points of the eastern portion of the Central Alps, the following species have been observed by Messrs. Schlagintweit up to 10,360 feet at different points : *Androsace glacialis, Cerastium alpinum* and *latifolium, Cherleria sedoides, Ranunculus glacialis, Saxifraga biflora, hypnoides, cæsia* and *oppositifolia* (this sometimes in great predominance over other species), and *Silene acaulis.*

On Pez Linard, in the Rhætic Alps, Heer observed *Androsace glacialis, Gentiana bavarica, Cerastium latifolium, Chrysanthemum alpinum, Ranunculus glacialis, Saxifraga bryoides* and *oppositifolia,* and *Silene acaulis* between 10,000 and 10,700 feet.

Saussure observed *Androsace glacialis* and *Ranunculus glacialis* on the Col de Mont Cervin, 10,461 feet ; *Aretia helvetica* (the only flowering plant) at 10,578 feet, on the Col du Géant ; and at 10,680 feet, on Mont Blanc, one *Silene acaulis.*

S

Zumstein found *Chrysanthemum alpinum* and *Phyteuma pauciflorum* by the Lys glacier on Monte Rosa, at 11,352 feet.

When we compare the arborescent vegetation of the Alps with that of the north of Europe, we perceive that the former undergoes a modification, as we pass from the external chains to the central masses, similar to that which the latter exhibits as we advance from the west coast into the interior of Europe.

This is shown by the appearance of the larch and *Pinus Cembra* in the higher Alps, trees which are wholly wanting in the mountains of the west of Europe, in the Pyrenees, and Scandinavia, while they occur on the Carpathians, and abound in Siberia. The predominance of the spruce over the Scotch fir is another evidence, since the former is much more favourably situated in a continental than a coast climate, as is seen in Scandinavia. A further indication is given by the low limit of the beech in the central mountains; for this tree, attaining its greatest development in the lowlands bordering on the North Sea, is arrested in Sweden even south of the limit of grain-culture, and is not found anywhere in the east of Europe, except in the south of Russia. Moreover, the comparatively high level of the limit of the vine may be mentioned, this plant being more suited to a continental than a coast climate.

As with the trees just mentioned, the vegetation of the cereal plants is favoured by the continental climate; since the severity or mildness of the winter does not affect them importantly, while the heat of summer in the interior of Russia allows of their culture in latitudes having a lower mean annual temperature than those which mark the boundary in Western Europe. The corn-plants are particularly remarkable for their high limit on the Central Alps.

We should be led to suppose from the foregoing facts, that the climatal conditions of the high Alps and the lower ridges bore the same relation as the interior of the continent does to the western coast. But with regard to temperature this is certainly not the case, for with the increasing height the summer heat diminishes, and thus approaches nearer to the winter cold, causing a more equable climate than in the plains. Such being the case, it may be supposed that the great intensity of the sun's light upon the higher Alps acts in some degree as an equivalent for the lower temperature, since this intensity of light, unaccompanied by elevation of the temperature of the air, is one of the most striking distinguishing meteorological conditions of mountains.

The condition of moisture of the air and soil are important elements in this question. Here the conditions appear at first sight contradictory, for the absolute annual amount of rain increases with the height, tending to produce the coast climate ; but on the other hand, the proportion of the summer rains to those of autumn and winter rains increases also with the height, as does also the absolute dryness of the air at any given time, two conditions belonging to the continental climate. Much of the richness of the vegetation of the Alps must be attributed to these conditions. The damp soil, watered by frequent rains and melting snow, affords abundant nourishment to the roots, while the dry air and intense sunlight stimulate the development of the aërial portion of the plants. Where the sources of moisture are more scanty, as on the more continental Carpathians and on Etna, neither high enough to maintain a constant reservoir of snow, the uppermost tracts are characterized by great barrenness, attributable probably in great part to the dryness of the atmosphere.

A comparison of the Alps with the Scandinavian mountains affords some strong distinctions. The latter have a greater extent, but the former are twice as high, both as regards the mean elevation and greatest altitude, Mont Blanc being 14,800 and Skagestoltind 7000 feet. The direction of the Scandinavian mountains is nearly N. and S., the Alps rather E. and W.; the former stretch through 13°, the latter through $4\frac{1}{2}°$ of latitude; thus the former afford a greater variety of conditions of climate and vegetation at their base; but as the Alps attain so much greater height, they do in fact surpass the Scandinavian mountains in variety. The latter are more open to the influence of the ocean, and the sea penetrates them in deep bays; the west side thus exhibits a perfectly insular climate. This is not the case in the Alps. The Scandinavian mountains are very abrupt in the west, and sink very gradually to the east; there is not such a striking contrast between the N. and S. faces of the Alps. The Scandinavian mountains are flat-topped, and the passes are over mountain plateaux; in the Alps there are no such elevated plains, and the passes go through deep depressions between the great masses, where valleys running in opposite directions intersect. The Alps have great longitudinal valleys; these are unusual in Scandinavia: the cross valleys on the opposite sides of the Alps are not so different as in the northern mountains; and the same is true of the rivers, for these are large in the E. and comparatively small in the W. in Scandinavia, while the difference is much slighter in the Alps, although the northern declivity has more considerable streams than the southern. There are large lakes at the foot of the Alps on both sides, in Scandinavia only on the east; in the latter are found large alpine lakes, in the former only small ones. Limestone, which abounds in the Alps, is wanting in the Scandinavian mountains.

The snow-line on the north side of the Alps is 6000 feet higher than in the northern parts of Scandinavia, and on the south 3400 feet higher than in the southern confines of Scandinavia; but from their greater elevation the Alps have more extensive masses of snow and ice.

In Scandinavia the tree-limit is indicated by the birch, in the Alps by firs. The two lower mountain zones of the Alps, the regions of the beech and the chestnut, do not exist in the Scandinavian mountains. Compared with the climate and tree-limits, the cultivation of corn does not go so high in the Alps as it does toward the north; for it ceases about with the beech in the Alps, and grazing is the regular pursuit in the region of firs; while in Scandinavia the beech only goes to 59°, and corn-culture to 70°, that is, as far as the Conifers. Corn succeeds in the latter under a mean temperature below the freezing-point, while in the Alps it ceases at 41° Fahr. The cause of this is the hot though short summer of the north. The Alps have maize and the vine, which will not grow around the Scandinavian mountains; the meadows are throughout richer in the Alps, and grazing is therefore much more extensively pursued.

SECT. 8.

The Spanish Peninsula.

THIS part of Europe is completely cut off by the narrow range of the Pyrenees, stretching in the direction E.S.E. and W.N.W. to a length of about 270 geographical miles from the Mediterranean to the Atlantic, having an average breadth of only twelve geographical miles, and lying between the $42\frac{1}{2}°$ and $43\frac{1}{2}°$ N.L. They are unconnected with the Cevennes or the Alps, are bounded on the north by the canal of Languedoc, and on the south by the valley of the Ebro.

The Pyrenees.

The Pyrenees constitute an immense natural barrier, separating France from Spain; and from the main ridge, extending from sea to sea, transverse ridges stretch out here and there like buttresses. The real western termination, indeed, is not at the Port des Passages, on the Bay of Biscay, for they are continued onwards almost to an equal length through the provinces of Biscay, Asturias and Gallicia, to Cape Finisterre. Here, however, we have merely to speak of that portion lying between France and Spain.

The mean height of the central and most elevated parts amounts to 7000 feet; the highest points are :—

	Feet.
Vignemale	10,971
Mont Perdu	11,137
Pic Posets	11,245
Pic Nethou (Maladetta)	11,392
Montcalm	10,633
Canigou	9,116

The Pyrenees really consist of two chains, one, the western, rising gradually from the Atlantic to the Maladetta, its highest point; and the other, or eastern, commencing with low hills to the north of the Maladetta, and increasing in height as it approaches the Mediterranean, near which lies Mont Canigou, one of its highest summits. From the point of dislocation, an important lateral ridge runs out on the north side through the department of the Hautes Pyrénées, of which the Pic du Midi de Bigorre is the principal peak. The loftiest summits of the Pyrenees are all out of the main chain; the highest of all, the Maladetta, lying to the southward, as also Mont Perdu, the next in altitude. The defiles (called " Ports" in the middle chain, and " Cols" in the transverse ridges) are all of considerable elevation, often from 7000 to 9000 feet, and there exist only two practicable roads for carriages, one at each extremity of the chain. The ascent is more abrupt on the Spanish side than on the French, where parallel ridges (at least two in number) are usually distinctly traceable. The Spanish Pyrenees are also watered by fewer streams, have fewer lakes, and are less clothed with forests than the French. The sides of the valleys are almost always steep, and the basins among the mountains are ordinarily small, and occupied by little lakes or deposits of alluvium; peat deposits are rare.

The elevation of the line of eternal snow is differently given by different authorities. Ramond states it to be situated at about 8100 to 8400 feet, but Mr. Spruce considers it to average 9000 feet, therefore about 1000 feet higher than on the Alps. It varies, however, with the exposure, and even the form of the mountains, and always descends lower on the eastern exposures than elsewhere. Since only the highest peaks exceed this line, there are no very extensive masses of eternal snow, and therefore few glaciers; and the snowy zone is never more than from 1000 to 2000 feet broad.

The Pyrenees are exceeded in height by the Alps, but are more lofty than the Scandinavian mountains, or any of those of Central Europe. They are exceeded in horizontal extent by the Alps, the Scandinavian mountains and the Carpathians.

The waters of the southern slopes almost all fall into the Ebro; on the north side, the western half is drained by the Adour, while the space from the source of the Adour to that of the Arriège is occupied by the head of the basin of the Garonne; on the extreme eastern angle a number of small streams run directly into the Mediterranean. The botanical districts, however, are not well defined by these limits, and the Pyrenees are more conveniently divided into three portions, the western, central, and eastern. The *central* district includes the upper part of the department of the Haute Garonne and most of the Hautes Pyrénées, in France, part of Arragon, and a very small angle of Catalonia, in Spain; it is drained by the upper parts of the branches of the Adour and Garonne, and contains the highest mountains and deepest valleys, as well as the most extensive forests. It contains the only glaciers of any extent, the principal of which are those on the north slopes of the Maladetta and Crabioules. The *western* district includes the department of the Basses Pyrénées and part of the Landes, besides a portion of the Hautes Pyrénées, in France, with a small part of Navarre, and the northern part of Arragon, in Spain. It extends further north than either of the others, and is therefore colder at the same altitudes; the sandy plains bordering on the Adour and the ocean have also a much more humid climate. The *eastern* district is contained between the *central* and the Mediterranean; in France it occupies the departments of Arriège and the Pyrénées Orientales; in Spain, nearly all the northern part of Catalonia. This district is the most southern, the warmest and driest, and also the most denuded of forests.

Clay-slate is the most important rock entering into the composition of the Pyrenees; and transition limestone constitutes a considerable proportion of the surface, especially in the west. Granite, gneiss, and other primitive rocks occur extensively, but do not form the highest summits, as in the Alps; they occur most extensively in the east. Some of the grandest features of the Pyrenees are formed by escarpments of oolitic limestone.

The following is an imperfect table of temperatures :—

	Lat.	Elev.	Ann.	Winter.	Summer.
Perpignan	$42\frac{1}{2}$	..	60	$45\frac{1}{2}$	75
Dax	$43\frac{1}{2}$	..	57	44	69
Mont Louis ..	$42\frac{1}{2}$	5195 ft.	$43\frac{1}{2}$	$31\frac{1}{2}$	57

According to these means, the climate of the north side appears to be milder at the Mediterranean than at the Atlantic end, and the difference of the seasons less at the latter, as would be expected. The mean temperature of Mont Louis, at an elevation of 5000 feet, only slightly exceeds that of Stockholm; but the winter is milder and the summer cooler. The vegetation of the Pyrenees is exceedingly rich and varied, and is composed of plants partly common to them and the mountains of Northern Europe, partly of plants common to the Pyrenees and Alps, partly of plants common to the Pyrenees and the mountains of Spain generally, and a certain number of peculiar species, together with a quantity of the common plants of Central Europe.

In tracing the distribution according to elevation, a number of zones have been made out. From the plains, which are spoken of under the head of the adjacent districts, seven successive regions may be distinguished, namely:—1. The *lower hilly zone*, which may be said to extend to the limit of the cultivation of the vine; and this is pretty nearly identical with the extent of the chestnut woods, so prevalent in the Spanish peninsula. 2. To this succeeds a *middle hilly zone*, defined above by the cessation of the culture of maize, at which point the box-tree begins to flourish luxuriantly; among the characteristic plants of this zone may be mentioned *Saxifraga Geum, Asperula cynanchica, Euphorbia hyberna* and *dulcis, Cephalanthera ensifolia*, &c. 3. From the limit of the vine to about 4200 feet extends the *upper hilly zone*, in which the cultivation of rye, potatoes, cabbages, &c., is carried on, ceasing at its upper limit. The box here flourishes luxuriantly, and is accompanied by *Saxifraga Geum, Heracleum pyrenaicum, Arnica montana, Cirsium monspessulanum, Teucrium pyrenaicum, Scrophularia Scopolii, Erinus alpinus, Calamintha sylvatica*, &c. The three preceding zones are about equal in breadth. 4. The *subalpine zone* extends from the limits of the cultivation of esculent vegetables, at about 4200 feet, to that of the spruce fir, the beech extending up to within a few hundred feet of the upper limit. The vegetation of this zone is composed of mountain plants mingled with certain species common in the plains, and the true subalpine species here attain their fullest development, both in size and number. Meadows are scarce in this zone, and do not occur above it. Among the characteristic plants may be noted : *Meconopsis cambrica, Hutchinsia alpina, Viola cornuta, Dianthus monspessulanus, Saponaria ocymoides, Saxifraga Geum* and *aquatica, Chærophyllum hirsutum, Sambucus racemosa, Ramondia pyrenaica, Teucrium Chamædrys, Tozzia alpina, Nigritella angustifolia, Lilium pyrenaicum,* &c.

The distribution of the trees in these four zones may be roughly given as follows :—the chestnut extends from the plains to about 1400 feet ; the oak, from the plains to about 5000 feet ; the beech, from 2000 to nearly 6000 feet ; the spruce fir and the yew occupy a zone from about 4600 to 6000 feet ; the birch, which is so common in the Alps and the Scandinavian mountains, occurring in a zone above the oak, is excessively rare in the Pyrenees, if not altogether wanting. In the Alps, also, there is no tree above the spruce fir ; but in the Pyrenees this is not the case, for we find the Scotch fir in the zone next to be mentioned.

5. The *infer-alpine zone* extends from the limit of the spruce fir at 6000 feet to about 7200 feet, and is chiefly characterized by the presence of the Scotch fir, which, however, seldom passes the upper limit even in the most stunted form. *Rhododendron ferrugineum* is lost in this zone at about 6600–6900 feet, and above this altitude the herbage is chiefly composed of *Nardus stricta* and *Festuca eskia (varia γ. crassifolia,* Koch). Low shrubs occur here, such as *Vaccinium Myrtillus* and *uliginosum, Empetrum nigrum, Sorbus Chamæmespilus* and *Salix pyrenaica. Crocus multifidus,* a conspicuous ornament of the lower mountains, reaches the very summit of this zone. *Silene ciliata, Arenaria ciliata, Helianthemum*

œlandicum, Viola biflora, Trifolium alpinum, Dryas octo-petala, Geum pyrenaicum, Epilobium alpinum, Saxifraga aizoon minor, Eryngium Bourgati, Aster alpinus, Vero-nica aphylla, Bartsia alpina, Pinguicula grandiflora, Androsace carnea and *villosa, Primula integrifolia, Glo-bularia nudicaulis* and *rupestris, Salix pyrenaica* and *re-ticulata,* &c. &c. are characteristic plants of this zone.

6. The *middle alpine zone* extends from 7200 to 8400 feet. *Festuca eskia* attains the upper limit in this zone, but *Nardus stricta* fails below it. The dwarf juniper is the giant of the contracted vegetation. Here the weeds which follow man and domesticated animals from the plains cease to appear. The following species are abun-dant in this zone :—*Statice alpina, Gentiana alpina, Po-tentilla nivalis, Cherleria sedoides, Silene acaulis, Iberis spathulata,* and *Pyrethrum alpinum.* More character-istic species are : *Draba aizoides, Ranunculus pyrenæus, Saxifraga arctioides, bryoides* and *muscoides, Asperula hirta, Gnaphalium Leontopodium* and *supinum, Campa-nula pusilla, Phyteuma hemisphæricum, Pedicularis py-renaica* and *rostrata, Soldanella alpina, Daphne Cneo-rum, Veronica alpina, Juncus trifidus,* &c. &c.

7. Lastly, the *upper alpine zone,* above 8400 feet, is characterized by merely adding to the plants of the pre-ceding zone a very small number of herbaceous plants, all *perennial,* rarely descending into the middle alpine zone. Such are : *Ranunculus glacialis* and *parnassifo-lius, Draba nivalis, Stellaria cerastoides, Androsace alpina, Sibbaldia procumbens, Saxifraga grœnlandica* and *androsacea.*

The most striking points noticeable in the vegetation of the Pyrenees, are the influence of the oceanic climate in the western portions, and the mixed character of the flora generally, depending on the conjunction of the arctic European species and those of the southern mountains of the peninsula, with the considerable number of endemic species.

The Spanish Peninsula.

Lying between the 36° and 43½° N. L., this region is bounded by the Pyrenees, the Mediterranean, and the Atlantic, being of rhomboidal form and without any deep indentations of the coast. To the east lies the little group of the Balearic isles. The greatest extent from N. to S. is 120, and from W. to E. 135 geographical miles.

The greater portion of the surface rises to a considerable height above the sea, and forms an elevated table-land, the mean altitude of which amounts to 2000 feet; Madrid lies on this plateau; Granada is still more elevated (2560 feet). Travelling from the N. or E. coast towards Madrid a considerable ascent is made, but the land does not sink again toward the interior from the edge of the highland. The ascent from the west is more gradual. Upon this table-land are situated several mountain masses, namely, a northern and a southern boundary ridge running from east to west, and three other chains stretching in the N.E. and S.W. direction. These five chains are in reality composed of many smaller ones, bearing distinct names, but they are here grouped according to general direction, and may be described under collective names.

The range forming the northern boundary of the table-land between the Bay of Biscay and the Valley of the Douro, is the Asturio-Gallician. They might be regarded as a prolongation of the Pyrenees, for they have the same general direction, and are not separated by any considerable depression. They appear to attain a considerable height; the highest point is called Pennas d'Europa.

The second range, Guadarama, lies between the valleys of the Douro and the Tagus; the western portion is known as the Sierra d'Estrella. The highest point, Penalura, attains 8198 feet; the Pas d'Ildefonso, 5119 feet; the royal palace of St. Ildefonso being situated at an elevation of 4000 feet.

The third range, Sierra Guadeloupe, between the Tagus and the Guadiana, and the fourth, Sierra Morena, between the Guadiana and the Guadalquivir, are much lower, both as to mean height and the peaks.

The Sierra Nevada, however, forming the southern boundary of the Spanish plateau, attains very considerable height. The loftiest peaks are the Cerro de Mulhacen, 11,464 feet, exceeding the highest part of the Pyrenees. The Albujarras, 9243 feet above the sea, project as a promontory of the Sierra Nevada toward the neighbourhood of the sea.

On the east side of the plateau there is no continuous boundary of mountains, only smaller isolated masses.

Detached from the table-land lies the Sierra Monchique, a longish chain on the south coast, the highest mountain of which, Sierra di Foja, runs to 4069 feet. Montserrat also, on the south of the Pyrenees, lies separate and attains an equal altitude.

The low plains of the Spanish peninsula are of small extent; the longest of them are those of Catalonia and Arragon, the strip of coast of Valencia, and the central portion of Portugal.

The islands of Majorca and Minorca are mountainous; in the former Silla Torellos is 5100, in the latter Monte Toro 4780 feet above the sea.

Since the north and south margins of the table-land lie in the vicinity of the coast, the rivers are small on their sides; and as the plateau falls principally towards the west, the largest rivers, the Minho, the Douro, the Tagus, the Guadiana, and the Guadalquivir, flow in that direction, into the Atlantic. The largest rivers flowing into the Mediterranean are the Ebro, the Guadalaviar, the Xucar, and the Segura.

Most of the Spanish mountains are composed of primitive rocks; the table-land of red sandstone, which gives the general surface an arid, parched aspect. There are no active volcanos, but basalt occurs in certain spots.

The climate has not been accurately observed, on a large portion of the peninsula; but those of Lisbon and Madrid are known, and afford a striking contrast between the W. coast and the table-land.

	Lat.	Elev.	Ann.	Winter.	Summer.
Lisbon	$38\frac{1}{2}$	. .	62	$52\frac{1}{2}$	71
Madrid.	$40\frac{1}{2}$	2000 ft.	59	$43\frac{1}{2}$	77
Gibraltar	36	. .	68	59	77

When it is remembered that Madrid lies not only about $2°$ north of Lisbon, but also 2000 feet higher, the mean temperature of the former appears very high, probably an effect of the heating of the vast plain. With regard to the seasons, the winter of Madrid is $9°$ colder, the summer $6°$ warmer; the climate thus appears comparatively more continental. The cold sometimes falls to $16°$, perhaps even lower, and the summer heat sometimes rises to $104°$; while in Lisbon the highest temperature amounts to about $102°$, the lowest to $27°$. Snow and frost are very rare in Lisbon, and not uncommon in Madrid.

The annual quantity of rain amounts to 28 inches in Lisbon, only to $9\frac{1}{2}$ in Madrid, and the table-land generally suffers from drought. The plateau and the west coast thus contrast strongly. In both, however, the summer rains are very slight, autumn and winter being the proper rainy seasons; thus only $\frac{1}{30}$th of the annual rain falls in summer, $\frac{1}{3}$rd in winter, in Lisbon. Estremadura is sometimes without rain for from seven to nine months. The air is uninterruptedly bright and serene in the table-land during the summer.

Eternal snow is met with only on the Sierra Nevada (hence its name); but the plateaux of the Sierras Guadarama and Estrella are covered by snow for four or five months in the year, while in summer they are used as pastures.

The coast provinces of the Spanish peninsula share the character of the other coasts of the Mediterranean, especially determined by the quantity of evergreen trees and shrubs; which, requiring a mild winter, do not occur upon the north side of the great mountain-chains separating Southern from Northern Europe. Among the most remarkable plants are the cork oak (*Quercus Suber*), the evergreen oak (*Q. Ilex*), and several other oaks which do not lose their leaves in winter as ours do; the laurel (*Laurus nobilis*), the myrtle (*Myrtus communis*), the lentisk (*Pistachia lentiscus*), the strawberry-tree (*Arbutus Unedo*), and the southern Conifers, the stone pine (*P. Pinea*) and the cypress (*Cupressus sempervirens*). Aromatic shrubs and herbs are much more frequent and abundant than in the north, particularly those belonging to the Labiate family, such as sages (*Salvia*), rosemary, thymes and savories (*Satureia*); with numbers of bulbous plants with showy flowers, such as hyacinths, *Nar-*

cissi, &c. A few plants of tropical forms also occur here ; thus the palms are represented by the dwarf palm (*Chamærops humilis*). On the other hand, there is no verdant turf like that which carpets the soil of the north of Europe.

The elevated plains, through their dry soil and the deficiency of rain, support but a poor vegetation, and are almost devoid of woods. The uncultivated tracts are covered with heaths, overgrown with *Cisti* and brooms. Many parts of Spain are now in this barren condition, where in former times, under the Romans and the Moors, a rich and productive agriculture thrived. This was the case with Estremadura, which now displays vast tracts of heath rarely disturbed by the foot of man, and where habitations are but sparingly scattered over the country.

Where the mountains are clothed with forests, which only occur extensively on the Asturio-Galician mountains and the Sierra Nevada, they are chiefly composed of the chestnut and the northern deciduous oak ; but in the more elevated portions the beech, Scotch and spruce fir, birch and ash, are met with.

The corn-crops principally raised are of wheat and maize ; rice is grown in Valentia and some of the plains of the western coast ; rye here and there on the northern highlands ; doura (*Sorghum vulgare*) is also cultivated, particularly in the west ; but the beans of the St. John's bread-tree, called ' garrobas,' are very extensively used as an article of food by the lower classes all over Spain. The chief fodder plants are barley and oats.

The culture of the vine is almost universal, and the products of very superior quality, of which it is scarcely necessary to mention the names of Oporto, Xeres, and Malaga. The olive is much cultivated in the coast pro-

T

vinces, but sparingly on the higher grounds. The various plants of the orange tribe flourish richly on the coasts, and grow even in Galicia, but do not bear the climate of the plateau. The other southern fruits, figs, almonds, peaches, apricots, &c., are produced in profusion, and of excellent quality; the two first, with raisins, being very extensively exported. In the south of the Peninsula the cotton-plant (*Gossypium herbaceum*) is grown, and even the sugar-cane thrives. The date-palm bears fruit, which, however, scarcely ripens properly. The culture of the mulberry (*Morus alba*), with the production of silk, is of some importance. More detailed accounts of these points will find a place in the following sketches of some of the most characteristic and best-known provinces of the Peninsula.

The vegetation of the extreme north of the Spanish peninsula differs very much from that of Andalusia, and by way of contrast to the description of that region, presently to be given, we select the provinces of Guipuscoa and Biscay, concerning which we have most information.

Guipuscoa and Biscay lie in the same latitude as Provence, a part of France which partakes of the Mediterranean flora, but the great difference in the character of their vegetation affords a striking example of the effects of local influences. For Provence is sheltered from the north winds by the lofty Alps; it lies fully open to the warm breezes continually flowing up from the coast of Africa, while on the northern littoral of Spain just the opposite conditions occur; the Cantabrian chain, the western prolongation of the Pyrenees, a broad and rugged range of a mean height of 5000 feet, separates, like a wall, the northern coast of the peninsula from the hot plains of the interior, and cooperating with the snowy peaks of the Pyrenees on the E.S.E., considerably cools down the warm air from the Mediterranean before it reaches this region; at the same time that the coast itself is washed by the great ocean and lies fully open to the northerly winds and to the raging storms of the Bay of Biscay.

The provinces of Guipuscoa and Biscay form a romantic mountainous country, the highlands being for the most part ramifications of the western Pyrenees and their prolongations. The mountains, which are mostly composed of limestones and sandstones, contain a great abundance of springs, whence brooks and little rivers traverse the country in all directions. The temperature is very equable, seldom high, and hardly ever falling below the freezing point, while rain falls almost daily. The quantity of streams, together with the frequent rains, keep both the atmosphere and soil constantly moist, and hence the mountains, valleys and plains of this region are almost uninterruptedly clothed with a luxuriant vegetation. The moisture is the more necessary to this, that the soil is very thin even in the valleys; the declivities of the mountains are ordinarily rocky, the summits usually naked peaks of stone.

The vegetation of these provinces bears a much more striking resemblance to that of the Rhine provinces and S. Germany, that is to say, to that of Central Europe, than to the vegetation of the South. The mountains are not bare, as is commonly the case in the south, but the declivities, at least, are more or less thickly clothed either with forests or bush. The forests are mostly composed of young trees, because the old woods were almost destroyed in the long wars which have raged in these provinces; they consist exclusively of deciduous trees, in the lower part of the chestnut and the common oak (*Q. pe-dunculata*), higher up of another species of oak and of beech. The bush, or *montebaxo*, is very complex in its constitution in the lower grounds, but in the higher tracts, particularly where it covers the crests and upper parts of the slopes of the mountain ridges, it principally consists of the common furze and various species of heath.

The lower parts of the slopes in the valleys and low grounds are carefully cultivated, in fields of rye, wheat and clover, with vineyards and orchards; while among them are seen rich grassy meadows, such as are never met with in the south of Europe. The vegetation of these meadows is almost wholly like that of the meadows of the north; southern forms do indeed appear here and there, but these fields, with their yellow buttercups, red clover, blue-bells and forget-me-nots, together with the many apple, pear, plum, and cherry trees, and the oak and beech woods, stamp the resemblance of the landscape to the northern type.

The vicinity to the south is detected most clearly in the vegetation of the hedges, on the walls, and in the *montebaxo* of the hills and slopes; the flora of the mountains is a mixture of the plants of Central Europe with those of the Pyrenees. Very few north-Spanish, or as they are termed by Mr. Bentham, *oceanic* plants, have been observed. The vegetation of Guipuscoa and Biscay is therefore a compound of Central and South European plants, with the intermixture of a few Pyrenean and oceanic species. The vegetation, consequently, like the climate, the contours of the landscape, and the geographical position, is neither decidedly northern or southern, but forms a link or transition state, between the two.

This transitional character explains the fact that no

family or genus is greatly predominant in number of
species. The species most abundant in absolute number
are those trees and shrubs already mentioned, together
with grasses and ferns. It is remarkable that the Cruciferæ are almost wanting here, while the common juniper
is the only indigenous representative of the Conifers.
The Cistaceæ, too, which seem to have their true home
in the Spanish peninsula, occurring so abundantly and in
so many regions, are almost entirely absent from this
province, a fact the more remarkable, since various *Heli-
anthema* are common in the west of France; they were
only found by M. Willkomn to the west of the Rio de
Bilbao; in Guipuscoa not a species was met with.

The evergreen oak (*Quercus Ilex*) is the most striking
representative of the south among the wild trees; in
cultivation the cypress, laurel and fig-tree are met with.
The evergreen oak occurs most abundantly in the valleys
and depressions of Biscay; it is rarer in Guipuscoa, but
nowhere does it attain the same size and beauty as in
the south of the peninsula, appearing either merely as
a shrub, forming an inconsiderable constituent of the
"bush," or as a poor tree. The cork oak does not seem
to grow here at all. Among the shrubs, the furze, although a northern species, contributes most to give a
southern physiognomy to the landscape, from its resemblance to the thorny brooms which form so large a proportion of the *montebaxo* of the south. The heaths,
which it is true are shrubs of some size, give more of the
aspect of the heathy tracts of N. Europe than that of
the southern bush-tracts, and one of them is actually a
northern species, namely *Erica Tetralix*. Among the
herbaceous plants, the Monocotyledons would appear to
give the most character, in *Serapias Lingua, S. cordi-*

gera, Asphodelus albus, Smilax aspera, &c.   The Pyrenean flora is represented most abundantly in these mountains, since lower down, *Aquilegia viscosa,* common in the meadows and hedge-rows, is perhaps the only example.   The "oceanic" type is abundantly represented by the beautiful *Menziesia polifolia,* which is frequent in hedges and bushy places, especially in the mountains separating Guipuscoa from Biscay.   The ferns, which are so conspicuous in these regions, favoured so greatly by the mild, moist climate, are all common English species, except *Adiantum capillus Veneris,* which, as with us, is among the rarest species, as *Pteris aquilina* is the commonest.

On the east side Guipuscoa is separated by a prolongation of the Pyrenees from the province of Navarre, and in this the vegetation begins to assume a totally different character.   It is a mountainous region, with plateaux, on one of which the city of Pamplona is situated. Here pine forests and *montebaxo,* formed principally of box, clothe the mountains; the furze gives place to a thorny broom; while in the valleys occur the Mediterranean *Anchusa italica, Sideritis hirsuta, Orchis fragrans,* &c.   It is not worth while to dwell upon this region, but for an example of the vegetation immediately on the south side of the Pyrenees we will pass to Arragon, lying still further east.

The name of High Arragon is applied to the mountainous region lying between the valley of the Ebro and the French border.   It therefore includes the south declivities of the Pyrenean chain, the numerous valleys of the central Pyrenees opening to the south, and also the chain of mountains running parallel with the Pyrenees, between them and the basin of the Ebro.   This chain, the highest summits of which attain more than 5000 feet, is usually regarded as a branch of the Pyrenees, but it is really quite detached from it; for, according to M. Willkomm, the numerous chains represented on maps as running from N. to S. from the Pyrenees have no existence, but two or three chains run parallel with the Pyrenees, that nearest to them, the highest, being separated completely by a broad valley. From the borders of Navarre to the town of Jaca, this constitutes the very wide basin of the Rio Aragon; to the east of Jaca it is an equally broad, but by no means mountainous plateau, traversed by the Rio Gallego, and extending far inwards to the east. Whether it reaches to Catalonia, M. Willkomm was not able to decide by actual investigation; but as the geological structure of the chain, consisting of a breccia composed of rolled fragments of the rocks of the Pyrenees, united by a sandy cement, is identical with that of Montserrat, it is probable that it does extend into Catalonia, and that the isolated mountain of Montserrat is a terminal, detached member of this chain, which commences with abrupt peaks on the borders of Navarre.

The close vicinity of the Pyrenees would lead us to imagine that the vegetation of this more southern chain is similar, and in fact the greatest portion of its plants belong to the Pyrenean flora; but there appear to be a certain number of plants peculiar to the *breccia*, since they have not yet been found either in the Central Pyrenees or the south side of the valley of the Ebro.

The soil of the valley of the Rio Aragon is composed of diluvial deposits from the Pyrenees and the opposite chain, forming irregular, earthy, marly or slaty hills at the foot of the mountains on each side; the distribution of wood is rather irregular. Oak woods occur on sandy soil, but Coniferous woods, of the Scotch fir, are the most abundant, clothing all the sandstone hills forming the extreme southern promontories of the Pyrenean chain. The slopes of the *breccia* mountains are closely covered either with forest or bush, the former composed of Scotch fir, oak, or the narrow-leaved ash; the bush, of box, maple, *Cratægus monogynus*, *Cytisus sessilifolius*, broom, roses, brambles, &c. About the hedges are found abundance of common Mediterranean and Central European plants. The Peña de Oroel seems to be the highest point of the breccia chain; on the south and south-east sides its slope is gradual, on the opposite rather steep. With the exception of the precipices of the west, the broad base, and the upper part of the southern slope, the whole mountain is densely clothed with pine forests. The broad base is in great part covered with bush, consisting of box, holly, *Quercus Ilex*, broom, &c., with lavender, *Aphyllanthes monspeliensis*, &c. The forest commences at about 3000 feet, and is made up of *Pinus pyrenaica*, with a good deal of *Abies pectinata* intermingled above. With these the mountain ash, *Cratægus*, *Amelanchier*, *Fraxinus angustifolia*, the holly (often a tree), dog rose, and the beech as a shrub. The soil of this shady wood is usually covered with a grassy turf, on which are found many common N. German plants, while more Pyrenean species occur on the barer rock and cliffs, in the upper part taking on a subalpine character.

The valley of the Gallego, on the south of this chain, resembles the valley of the Arragon; and the chains which separate this from the basin of the Ebro, composed of limestones and sandstones, have a vegetation analogous to that of the *breccia* chain, excepting that from their less elevation they are without the subalpine flora, and the most abundant plant in the " bush" which clothes them is the gall oak, *Q. coccifera.*

Of the north-western corner of the Peninsula comparatively little is known; we are acquainted with it more from the accounts of general travellers than of botanists, none of whom appear to have penetrated into it in recent times.

The climate of Galicia is temperate and rainy, the surface being very mountainous. The mountains, as already noticed, are a prolongation of the Pyrenees from the Basque province to Cape Finisterre. The hills are well covered with timber, oak and chestnut; the meadows green and luxuriant; the banks of the rivers have quite an English aspect. The principal cultures in the higher regions are maize, rye and flax, with the orchard fruits, apples, pears and nuts, together with the gooseberry, currant, &c., which are rare in the hotter parts of Spain.

As the eastern mountain boundary is covered almost all the year with snow, while the sea-coasts and river-valleys, lying in a latitude of 42°, have scarcely any winter, the range of vegetation is very wide and interesting; but it has never been thoroughly investigated yet. The warmer and lower valleys of the Minho, and the country about Tuy, Redondela and Oreuse, are described as perfect gardens, the olive and orange flourishing, and rich wines being produced, which would rival those of Portugal if more care were devoted to the culture of the vine.

Speaking of the shores of Vigo Bay, Mr. J. Ball says, that the aspect of the spontaneous vegetation at this extremity of the Peninsula is certainly calculated to confirm the opinion of those naturalists who believe in the former extension of the Galician coast toward the British Islands; but as far as the evidence goes, it points to a connexion with Cornwall and Devon rather than to any special relation with the flora of the west of Ireland. The majority of the species seen by Mr. Ball in a brief walk were either common throughout Western Europe, or plants which in Britain are characteristic of the S.W. of England. Among the first were *Ulex europæus, Thrincia hirta, Centaurea nigrescens, Veronica polita, Calamintha Nepeta, Polygonum Hydropiper* and *laxum*, &c. Of the second class, the following, with the exception of the first, which extends to Ireland, are in Britain peculiar to the S.W. of England :—*Coronopus didyma, Oxalis corniculata, Corrigiola littoralis, Polycarpon tetraphyllum, Scrophularia Scorodonia, Mentha rotundifolia,* and *Asplenium lanceolatum.*

Not more than seven or eight species were observed which do not belong to the British flora, and this circumstance is that which most strongly tends to confirm the belief in the existence of something more than an accidental similarity arising from climatal conditions. Among these exceptional species are named :—*Conyza ambigua, Andryala sinuata, Oplismenus Crus-galli, Euphorbia pubescens,* and *Chenopodium ambrosioides;* the first three of these are widely spread throughout the warmer temperate zone, and their presence does not indicate any peculiarity in the flora.

The country about Valentia has an extremely rich vegetation; blest with a delightful climate, the activity of the inhabitants renders the vicinity of this city a wonder to every foreigner. Water-courses, with water-wheels, are seen in every direction; the fields of wheat and hemp are surrounded by rows of mulberry-trees, while on the south and east are seen large plantations of olives, which are much larger and more beautiful than the dwarf olives of Provence. Figs, citrons and orange-trees grace the country-houses; but the date-palm is rather rare, although growing to a height of from 40 to 60 feet. Rice is largely cultivated near the lake of Albufera, and the water-courses are everywhere filled with flags and reeds. Near the sea grow the tamarisk (*T. gallica*), with the American aloe and the cactus (*C. Opuntia*), which is cultivated in gardens for feeding cochineal, but grows wild on every sunny, stony spot in the maritime region; about the castle of Murviedro it forms impenetrable thickets five or six feet high, and its stems are often five or six inches in diameter.

Near Albufera occurs a wood of the Aleppo pine (*P. halepensis*), which contains abundance of the original plants of the district; in such places the underwood consists of the gall-oak (*Querc. coccifera*), myrtle and dwarf palm (*Chamærops*), and with these are found the lentisk, heaths (*Erica arborea*), juniper, rosemary, &c. The adjacent sandy hills on the sea-shore are covered with cistuses (*Cistus albidus* and *salvifolius*), also *Passerina hirsuta*, and the apple of the Dead Sea (*Solanum sodomæum*) with stems as thick as a man's arm.

The Sierra de Chiva, a chain of limestone mountains passing out from the great plateau nearly to the sea, is broad, and as much as 6000 feet high. It is intersected by deep cross valleys (called by the natives *Barrancos*), which were once covered with pine-forests; a few isolated trunks of the Aleppo pine are all that remain; the dry declivities, with very few springs, are now overgrown by the low bush or *montebaxo*, only the extreme summits being bare.

Three regions of Mediterranean vegetation have been distinguished upon these mountains, reaching up to 4000 feet; above that the vegetation assumes a North-European character.

1. From the sea-level to about 500 feet extend the American aloe and the cactus, and this is the region of the cultivation of the algaroba, or St. John's bread-tree (*Ceratonia*), so remarkable a plant in Spanish economy. The *montebaxo*, or bush, is here composed of the dwarf-palm, arborescent heath, *Daphne Gnidium*, brooms, furze, rosemary, and a few shrubby oaks.

2. From 500 to 2000 feet, that is, as far as the extreme limit of the dwarf palm, lentisk, *Juniperus Oxycedrus*, &c., the first-named plant and the rosemary predominate.   The *Erica arborea* is still met with, and *Rhamnus lycioides*, the pistachio (*Pist. terebinthus*), and certain Cistineæ are first seen here; the characteristic grasses being the feather-grass (*St. pennata*) and *Macrochloa tenacissima*, or *Esparto*, an important plant here, being woven into basket-work for mule-panniers, chairs, and the sandals worn all over Spain; and also worked into ropes made in great quantity in Marseilles.

3. From 2000 feet up to 4000 feet, the upper limit of the cultivation of wheat and the olive, the bush is composed of similar plants, associated with a new juniper (*Juniperus phœnicea*), a species of ash, the strawberry-tree (*Arbutus Unedo*), and the evergreen oak (*Quercus Ilex*).   The greater part of the slopes of this region are wild and uncultivated mountain-land.

4. From 4000 feet to the summits, at 5000–6000 feet, are found isolated pines and a bush composed of *Ulex australis* and *Juniperus phœnicea*, with many of the plants of Northern Europe, such as the common mallow, spurge, shepherd's-purse, dead-nettle, &c.

The only alpine region here is the summit of the highest mountain, Monte de la Santa Maria, on which grows the *Arbutus Uva-ursi*, with much yew, and a species of *Cotoneaster*. One Saxifrage alone was found by M. Willkomm, thriving in luxuriant tufts, and the asphodel (*A. ramosus*) covered the steep slopes; a pretty tulip, the grape hyacinth, and a species of thrift also occur there.

Throughout Valentia the country is very fertile, well-cultivated, and clothed with timber; the broad valleys frequently filled with rice-fields.

But on traversing the pass of Almansa into Murcia, this changes to an elevated, treeless desert and thinly-populated tract. Low uniform calcareous hills alternate with wheat-fields and waste sterile plains clothed with solitary specimens of umbelliferous plants and brooms.

> More bleak to view the hills at length recede,
> And, less luxuriant, smoother vales extend;
> Immense horizon-bounded plains succeed!
> Far as the eye can see withouten end,
> Spain's realms appear where'er her shepherds tend
> Flocks, whose rich fleece right well the trader knows.

In places the country is somewhat better, the low hills presenting little isolated woods of *Quercus Ilex* and plantations of dwarf olives; on reaching the other edge of the plateau, however, the broad green valley of the Tagus is seen, filled with abundance of trees. The woods about Aranjuez are remarkably fine. Gigantic planes, elms, limes, beeches and oaks clothe the banks of the Tagus for leagues, and among these are found many of the plants common in the chalk districts of England, such as *Chlora perfoliata,* &c., in company with the plants of the warmer parts of Central Europe.

The characters of the vegetation of Southern Spain have been well described by Boissier in an account of the coast-terraces between Gibraltar and Almeria, including the inland region as far as the highlands of Andalusia. Along the entire coast-line run a series of isolated mountain chains, the Serrania de Ronda (6000 feet), the Sierra Tejeda (6600 feet), and the Sierra Gador (7000 feet), which form the southern border of the plateau of Spain, since on the north side they pass directly, at an elevation of 2000 to 2500 feet, into the elevated plain of Ronda, the Vega in Granada, and the great plateaux of Guadix and Baza. The chain of the Sierra Nevada, which is narrow, but ascends in places to 11,000 feet, is interposed between the boundary chain and the plateau near Granada.

Many of the plants of this district are gregarious, and the flora includes more thorny species than any other country in Europe; hence bearing some resemblance to the steppes of Western Asia, although the families developing the thorns are not the same.

In the first place, however, we may devote a few lines to a description of the isolated point of Gibraltar, which has been thoroughly investigated, and presents several points of interest.

The Rock of Gibraltar, geologically belonging to the Jurassic system, rises abruptly from the isthmus which connects it with the mainland, attaining a height of 1439 feet. It is about $2\frac{3}{4}$ miles long and 1 mile broad. Excepting on the steepest cliffs the hill is covered with vegetation, so varied in its composition, that, including those of the isthmus, the flora of Gibraltar contains 456 indigenous species of flowering plants. The mean temperature of the town, from five years' observations, amounts to about 68° Fahr.; that of January to 58½°, that of July to about 79°. The greatest quantity of rain falls in January, November and December; February to April, and October, have a mean amount; the dry season lasting from May to September. The observations of twenty-five years afford an average of scarcely 70 rainy days annually. A very luxuriant development of vegetation begins to unfold itself even in December.

The vegetation of the Rock is composed chiefly of *montebaxo*; the more precipitous eastern side is covered with the dwarf palm, the young shoots of which furnish the favourite food of the native monkeys; elsewhere, the broomlike shrubs of Andalusia predominate, particularly *Genista linifolia* and *Sarothamnus bæticus*, with *Daphne Gnidium*. Many exotic plants have established themselves and become perfectly naturalized, such as *Oxalis cernua*, as in Malta, and *Phytolacca dioica*, which is planted to form avenues. The sandy isthmus affords a rich collection of plants; a considerable portion of it, however, is almost exclusively overgrown with *Cachrys pterochlæna*.

The plants composing the flora belong to various regions; thus about 50 species, or one-ninth, are of the endemic forms of South Spain and a part of Northern Africa; four species have as yet been found only near Gibraltar, namely *Iberis gibraltarica* and *Silene gibraltarica,* which are even absent in the environs and grow on this rock alone; *Ononis gibraltarica* is found only in the immediate neighbourhood, but *Cerastium gibraltaricum* has been found by Boissier in the Sierra d'Agua. A fifth species, *Cratægus maroccana*, very abundant on the Rock and not occurring elsewhere in Spain, seems to have come from Morocco. The rest of the *endemic* plants are natives of wider floral regions, mostly belonging to the general Mediterranean flora, but in part to the forms principally developed in Morocco; such as *Sempervivum arboreum, Eryngium ilicifolium, Cladanthus proliferus, Calendula incana* and *suffruticosa, Linaria lanigera, Stachys circinnata, Passerina tingitana, Ephedra altissima,* &c.

Passing to the mainland, we may distinguish four regions in South Granada, the first of which, the *warm region*, includes the slopes rising from the coast up to 2000 feet. Here intense atmospheric precipitation takes place in October and November; the spring rainy season is less regular and lasts through February and March, sometimes till April; uninterrupted drought prevails from April to September; thus the dry season is probably longer that at any other point in the region of the Mediterranean flora.

The vegetation passes through phases corresponding to this climate. After the dry season, Liliaceous plants are developed with the first rains of October or November, and these are succeeded by annuals which flower throughout the winter. April and May is the time when most plants are in flower; in June and July the annuals have withered, and herbaceous plants of the families Compositæ, Umbelliferæ and Labiatæ take their place; in August and September there is a profound repose of vegetable life, only a few species, such as *Mandragora, Atractylis gummifera,* and two or three Liliaceous plants, remaining. The most characteristic plant of the *warm region* is the dwarf palm (*Chamærops*), which covers large tracts and prevents cultivation. Here, as in Valentia, it only ascends to 2000 feet.

Among cultivated plants, the orange exactly characterises this region. In general, the soil is principally devoted to the culture of the grape, which ripens here at the end of August. Grain requires artificial irrigation; those places which can be reached by the water from the mountains, either through natural streams or aqueducts, sometimes exhibit most luxuriant fields of maize and wheat, shaded by orange and mulberry trees. But such oases are rare upon these bare and arid slopes; yet when present, the wheat is reaped as early as the latter part of June, and barley in May.

At the foot of these coast-chains exists a narrow, littoral district, sometimes flat and interspersed with salt lagunes, sometimes consisting of chains of low hills, and near Malaga alone forming an extensive alluvial plain. In this district alone (from the sea-level to 600 feet) are found the cultivated plants of the tropics, such as the sugar-cane, cotton, sweet potato, as well as the date-palm and algaroba; also the naturalized agaves and opuntias, as well as many indigenous plants, such as *Aloë perfoliata*, &c. The white poplar is the only native tree of this littoral district.

In the warm region as a whole 19 trees occur, among which a portion are, like the cultivated plants, of foreign origin. The only truly indigenous species are the algaroba, jujube, pomegranate, *Celtis australis*, and the white poplar, ascending to 2000 feet, and the following, which extend into the succeeding region, where they become more abundant : the fig (0–3000 feet; on the southern slope to 4000 feet), the olive (ditto), *Quercus Ballota* and *lusitanica* (—3000 feet), the cork oak (—4000

U

feet), *Quercus Ilex* (—4500 feet), and *Pinus Pinaster* (ditto).

Among the most important vegetable formations of this warm region are : 1. The 'bush' (*maquis* or *monte-baxo*), consisting of shrubs from 3 to 6 feet high, covering the greater part of the declivities, and consisting of the dwarf palm, many *Cisti*, mastics, numerous brooms, a few oaks, &c., under the shade of which abundance of annuals and grasses flower in winter and spring, the later perennials more rarely ; bushes of oleander decorate the damp margins of the streams. 2. The 'plains,' which are dry and bare, and on which aromatic undershrubs, such as lavender, &c., and especially *Centrophyllum arborescens* prevail. In other places the *Esparto* (*Macrochloa tenuissima*) occurs gregariously. In addition to these two principal formations there are the maritime plants of the littoral, the plants peculiar to the bare cliffs, and those indigenous to the swamps of Malaga. The entire region contains nearly 1100 species, among which those of the following families occur in the greater numbers :—Leguminosæ $\frac{1}{7}$, Compositæ $\frac{1}{8\frac{1}{2}}$, Grasses $\frac{1}{10}$, Cruciferæ and Umbelliferæ, Labiatæ and Caryophylleæ $\frac{1}{23}$, Chenopodeæ $\frac{1}{32}$, Scrophulariaceæ $\frac{1}{41}$, Cistaceæ and Boraginaceæ $\frac{1}{53}$.

The second, or *mountainous region*, characterized especially by the Cistaceæ, is that of the Spanish plateau, peculiar to this country and not admitting of a comparison with the vegetation of any other country of Europe. The peculiar condition is dependent on climatal causes which require some explanation.

In Italy, Dalmatia, and in Turkey, we find, immediately above the 'evergreen region,' wooded declivities clothed with the trees of Central Europe, and other plants indigenous on the north side of the Alps : the commencement of this central European region is often denoted, even at 1200 or 1500 feet, by trees which lose their leaves in winter.

In Spain, two 'evergreen regions' are distinguished ; a lower, which from the character of its vegetation appears to agree with the Italian or Dalmatian, reaching to 1500 feet in Catalonia and 2000 feet in Granada ; and an upper, which, extending from 2000 to above 4000 feet, includes the greater portion of Spain, and has no analogue in any other part of the South of Europe. Schouw's researches have shown that the climatal cause of the evergreen vegetation of the Mediterranean lies in the dryness of the summer, which the plants of N. Europe cannot endure. Out of Spain these last again find their natural conditions on the South-European mountains in the vicinity of the cloud region, where even in summer vapours are formed from the moisture of the atmosphere, producing the lower temperature of the northern climate. But the elevated plains of Spain are still drier, in summer, than the coast regions ; the damp, mild spring, which forces all the plants into blossom, is followed by a hot, dry summer, giving place to a cold winter ; the three seasons of the Russian steppes are thus distinctly represented.

This explains the reappearance of some plants of the Spanish plateau in the Crimea and in the elevated plains of Asia Minor ; yet the number is but small, for the contrast between the insular climate and the excessive one of the more inland continent is here so distinctly marked, that the majority of the plants of Spain are unable to endure the intense winter cold of the Oriental

u 2

plateaux and steppes.   A large proportion of the plants
of the Spanish plateau-flora is therefore necessarily composed of endemic plants, because no similar climatal conditions occur elsewhere in Europe.   But this is not so
strikingly evident in Granada as in Central Spain, since
the plateau character is less developed on the declivities
of the mountains and the vegetation is poorer in variety
of forms.   It is evident, however, that under such a
climate more plants of the evergreen coast-region can
occur than of North or Central Europe.

Boissier considers that the region from 2000 to 4500
feet, on the north, and 5000 feet on the south, of the
mountains of Granada, corresponds to the plateau of
Central Spain.   Within this range, not far from the
lower limit, lie the cities of Granada and Ronda, where
the thermometer regularly falls 6 or 8 degrees below the
freezing point for a few days in winter.   Near the upper
limit, as at the village of Trevelez, in the Albujarras, the
snow remains on the ground for four months, from December to April.   The summer temperature in Granada
is often higher than that of the coast, but the nocturnal
cooling is very considerable.   The distribution of atmospheric precipitations is like that of the lower region,
only that thunder-storms are not unusual during the
summer on the Sierra Nevada, and thus the soil seldom
becomes so completely parched as lower down.

The agriculture of this region is chiefly occupied with
wheat and maize, and both these plants are carried to its
upper limit.   Wheat is harvested in July, or at the more
elevated points in the beginning of August.   The culture
of fruit-trees extends up to the same altitude as that of
wheat ; the chestnut, mulberry and walnut to 5000 feet ;
pears and cherries somewhat higher ; the latter, in par-

ticular spots, as high as 6500 feet. The most remarkable circumstance, however, is the phænomenon that the olive and vine ascend almost to the same elevation; on the north slopes the olive to 3000 feet and the vine to 3500 feet; on the southern declivities both to 4200 feet, forming an exception to the general rule of relation of altitude to the extension northward at the sea-level, or, in other words, displaying a discrepancy between their vertical and horizontal areas.

The vegetable formations of this second region are almost identical with those of Castile; they consist of: 1. *Maquis* or 'bush,' of similar aspect to that of the region below, but formed in great part of different species. Brooms and *Cisti* are here more general, *Cistus populifolius, Genista hirsuta,* being especially conspicuous, together with *Sarothamnus arboreus, Ulex provincialis, Daphne Gnidium,* rosemary, &c.—2. *Open woods,* of the cluster pine (*Pinus Pinaster*), extending from 1200 to 4000 feet, and the Aleppo pine (*P. halepensis*), from 2000 to 3000 feet, or of evergreen oaks, such as *Quercus Ilex, Ballota,* the cork oak, &c. The underwood is composed of *Cistus* shrubs, which grow more thickly in proportion as the large trees are more scattered. In the Serrania de Ronda these woods are replaced by a mingled growth of the Pinsapo fir (*Abies Pinsapo*), ranging from 3500 to 6000 feet, and the alpine oak (*Q. alpestris*), from 3000 to 6000 feet. The following trees also occur in this region: the ash, from 3000 to 5000 feet; the elm (*U. campestris*), from 2000 to 4000 feet; the black poplar 2000 to 5000 feet, and the stone pine 3000 feet.— 3. *Tomillares,* consisting of low half-shrubby and herbaceous perennials of the families of Labiatæ, Compositæ and Cistaceæ, forming a dense coat of vegetation over the soil, and varied here and there by tall tufts of feather-grass. The characteristic forms here consist of species of thyme, sage, lavender, helianthemum, wormwood, &c., and the feather-grass *Stipa Lagascæ.*—4. *Meadows,* of harsh, tall grasses, little visited by cattle, composed chiefly of wiry oat-grasses ; *Festuca granatensis* and the Esparto clothe the isolated slopes.—5. *Thistle-vegetation* on the fallows of the clayey soil.—6. A *gypsum vegetation,* with salt-marsh plants, occurring most extensively upon the elevated plains of Guadix and Baza. The characteristic plants, which mostly have a glaucescent aspect, and in part have fleshy leaves, are : species of *Frankenia, Peganum, Statice, Atriplex* and *Salsola,* with *Juncus acutus, Ononis crassifolia,* &c. The species collected in this region amount to 698, about one-seventh of which go still higher. The principal families are : Compositæ $\frac{1}{7}$, Leguminosæ $\frac{1}{14}$, Labiatæ $\frac{1}{16}$, Cruciferæ $\frac{1}{17}$, Umbelliferæ $\frac{1}{17}$, Grasses $\frac{1}{19}$, Scrophulariaceæ $\frac{1}{26}$, Cistaceæ $\frac{1}{30}$.

The third, or Central European region from 4500 feet on the north side, 5000 feet on the south, to 6500 feet, is characterized by the Scotch fir (*P. sylvestris*). In this the earth is buried in snow during four months ; thus it corresponds to the region of the Conifers in Central and Northern Europe. The distribution of temperature is tolerably accordant with the maritime exposure ; the cold is not extreme in winter, and the thermometer never rises above 77° Fahr. in summer. The atmospheric precipitations are here spread all over the year, mists and storms keeping the surface moist and fresh in spring and throughout the summer, in a higher degree on the northern slopes than on the southern ; thus we find here all the conditions influencing vegetation in Cisalpine

Europe. Agriculture is carried on only in the gardens
attached to the shepherds' huts (*hatos*); potatoes and
rye are grown, the latter mostly up to 6300 feet only,
though in particular spots on the south declivities it is
carried up into the succeeding alpine region. There
are no permanent habitations, and the land is used as
pasture, but connected tracts of grassy turf are rare,
shrubs and thorny plants clothing the greater part of the
declivities, so that cattle do not find here the rich supply
of food of other mountain ranges. The principal vegetable formations are : 1. *Low shrub* of the common
broom and *Genista ramosissima*, with a dwarf oak (*Q.
Toza*), ascending to 6000 feet ; near the shepherds' huts
these are replaced by thickets of the dog-rose and berberry.—2. *Open woods* of the Scotch fir throughout the
region, but of less extent in the Sierra Nevada ; the
trees are only 20 or 30 feet high ; on the Serrania de
Ronda are found the *Pinsapo* woods mentioned in the
previous region, with solitary yew-trees (5–6000 feet).
The Sierra de las Almijarras, southward of the city of
Granada, is clothed with fir-woods up to the peaks on
certain places, so that the existing forests would appear
to be the remnants of a destroyed girdle of Coniferous
trees, formerly covering all these chains. In the fluviatile valleys of the Sierra Nevada, isolated groups of trees
present themselves, the wrecks of larger woods ; in these
are found the mountain ash, *Cotoneaster granatensis,
Adenocarpus decorticans, Acer opulifolius*, the ash, *Salix
Capræa*, and *Lonicera arborea*, which ascends higher
than any other tree.—3. A low growth of *thorny under-
shrubs*, corresponding in altitude to the first formation,
occurs in the calcareous soils, composed of *Erinacea
hispanica, Genista horrida, Astragalus creticus*, &c.—

The region also contains numerous rock-plants, especially on the limestone, and in the valleys are found marshy tracts about springs, with meadows of limited extent, and these are the stations in which the greatest quantity of Central-European species occur. The principal families here are the Compositæ, Leguminosæ, Grasses, Cruciferæ, Caryophylleæ, Labiatæ, Scrophulariaceæ and Umbelliferæ.

The fourth region, the *Alpine*, is divisible into two subregions : *a.* the *region of alpine shrubs,* 6500 to 8000 feet ; and *b.* the region of alpine perennial herbs and grasses, 8000 to 1100 feet. In *a.* the snow begins to lie upon the ground in September, and the last masses do not melt before June in the succeeding summer ; thus the climate resembles that of the alpine regions of the Swiss Alps. The principal formations are : 1. The *Piorno,* a broad and connected girdle of vegetation extending to 8000 feet, composed of *Genista aspalathoides,* here and there replaced by the dwarf juniper and savine ; this growth stretches downwards in places, like the rhododendrons, into the subjacent wooded region (5500 feet).—2. Meadows of rigid grasses upon steep slopes, in the intervals between the *Piorno* bush, composed of *Avena filifolia, Festuca granatensis* and *duriuscula,* with *Agrostis nevadensis.* The prevailing families are the same as in the last region ; the mosses also are abundant here. In Boissier's statistics this is united to the Coniferous regions ; together they yielded 422 species.

The second part of this region (*b.*) is situated between 8000 feet and the summits (11,000 feet), and here isolated patches of snow persist throughout the year ; a continuous covering lying upon the ground during at least eight months out of the twelve. The earth is of course kept moist throughout the summer by the melting of the snow. No shepherds' huts are met with here, although the cattle are sometimes driven up as high as this. The vegetation consists of perennial herbs and grasses. Only four low shrubs belong certainly to this region, and two of these, *Ptilotrichum spinosum* and *Salix hastata,* are extremely rare ; the other two, *Vaccinium uliginosum* and *Reseda complicata,* have the woody stem creeping along the surface of the ground. The alpine meadows are here called *Borreguiles,* and are clothed with a close fine turf of *Nardus stricta, Agrostis nevadensis, Festuca Halleri* and *duriuscula,* among which flourish *Leontodon, Ranunculi,* gentians and other alpine herbs. In places the cæspitose perennials overgrow and displace the turf; among these occur *Silene rupestris, Arenaria tetraquetra, Potentilla nevadensis, Artemisia granatensis,* and *Plantago nivalis.* The alpine rivulets arise from little lakes (also in one spot from a glacier), and in these damp hollows are found taller plants, such as *Eryngium glaciale, Carduus carlinoides,* the foxglove, &c. Some plants are peculiar to the stony drift, such as *Papaver pyrenaicum, Viola nevadensis,* &c. ; others to the firm cliff, *e. g. Androsace imbricata, Draba hispanica,* &c. This region has yielded 117 species, of which one-third occur in the subjacent ; the principal families are the Compositæ, Grasses, Cruciferæ, Caryophylleæ, Scrophulariaceæ, Ranunculaceæ and Gentianaceæ. Only 5 annual species occur, 3 biennial, and 109 perennial.

Some points relative to the botanical statistics of this region, as laid down by M. Boissier, are worthy of notice. The plants which have been found in the lowest region amount to nearly 1100 species, and more than half of them are annuals ; this is the only known place in which the annual plants surpass or even equal the perennial. About 200 of these species are endemic to Spain, or extend only to Barbary or Provence, and only 12 species are found also in the East of Europe ; the characteristic families in these are Cruciferæ, Leguminosæ, chiefly brooms, Compositæ, nearly half thistles, Scrophulariaceæ and Labiatæ ; about 770 species are distributed over a larger area, but confined to the coasts of the Mediterranean ; and about 200 are Central-European plants, mostly either roadside or marsh plants. The second region contains about 698 species, of which 220 are Spanish plants only extending to Provence, 9 reappearing in the East ; about 200 Mediterranean plants and 260 Central-European species. The same families prevail here as in the preceding, with the addition of Umbelliferæ. The third and the lower subregion of the fourth region are taken together by Boissier ; and in them he gathered 422 species; among these 182 were Spanish plants, 101 of which have not yet been found out of Granada ; the characteristic families among these are: Cruciferæ, Caryophyllaceæ, Leguminosæ, Compositæ, Scrophulariaceæ and Labiatæ ; 185 species are Central-European, and only 55 Mediterranean plants. In the upper Alpine region 117 species have been met with : 45 are Spanish, 30 of which are as yet peculiar to the Sierra Nevada, 13 indigenous also in the Pyrenees ; 66 are alpine plants, in part occurring also on the plains of Northern Europe ; 6 species are common to the Sierra Nevada and the other mountains of Southern Europe.

The total number of species gathered in the entire district amounts to 2307 ; among which may be distinguished the following prevailing forms :—Compositæ $\frac{1}{9\frac{1}{2}}$, of which about $\frac{1}{3}$ are thistles (Cynareæ), Leguminosæ $\frac{1}{11\frac{1}{2}}$, Grasses $\frac{1}{14}$, Cruciferæ $\frac{1}{22}$, Umbelliferæ $\frac{1}{23\frac{1}{2}}$, Labiatæ $\frac{1}{24}$, Caryophyllaceæ $\frac{1}{25\frac{1}{2}}$ (Sileneæ $\frac{4}{9}$, Alsineæ $\frac{3}{9}$, Paronychiæ $\frac{2}{9}$), Scrophulariaceæ $\frac{1}{36\frac{1}{2}}$, Cistaceæ $\frac{1}{60\frac{1}{2}}$, Ranunculaceæ $\frac{1}{60\frac{1}{2}}$, Rubiaceæ $\frac{1}{62}$, Boraginaceæ $\frac{1}{64}$, Chenopodeæ $\frac{1}{68}$, Rosaceæ $\frac{1}{70}$, Liliaceæ $\frac{1}{70}$, Cyperaceæ $\frac{1}{72}$, and Orchidaceæ $\frac{1}{77}$.

The Sierra Morena is an intermediate chain of mountains between the coast ridge just described and the plateau of Castile. It extends from Murcia to Algarve, but is on the average only about thirty-six miles broad; the crests, for the most part, are about 2000–3000 feet high, the highest summits scarcely 5000. Its vegetation is remarkably uniform, the dense forests and tall shadowing shrubs forming a continuous tract of green and active vegetation, contrasting with that of the other mountains of Andalusia, which have only isolated patches of wood, and a low and barren 'montebaxo' or 'bush.' The principal formation of the Sierra Morena is grauwacke, occurring in gently rounded mountains and undulating ridges, alternating at Almaden with clay-slate, in the province of Huelva with gneiss, and southwards, near the low plain of the Guadalquivir, enclosed by other sandstone formations. The central portion is interrupted by the vast granitic formation of Cordova, constituting an elevated plain inclining from Hinojosa towards the north, and becoming connected there with white quartzose rocks, which, between Almaden and Fuencaliente, appear to form the highest chain of the whole Sierra.

The characters of the vegetation appear to be dependent on these geological variations. The prevailing shrub over the grauwacke, as far as Portugal, is the *Cistus ladaniferns*, which extends for more than 230 miles over the Sierra Morena, and frequently almost exclusively covers whole square leagues. Next in frequency to this are *Phillyrea angustifolia*, rosemary, and a *Helianthemum*. The forests on the grauwacke are composed of evergreen oaks (*Quercus Ilex* and *Ballota*) and the cork oak, the first mostly appearing only as a shrub.

The arid granite plateau is densely populated although sterile, but still possesses extensive woods of *Quercus Ilex* and *Ballota*,—the former, however, dwarf,—mixed with *Cistus ladaniferus, Phillyrea*, and the *Arbutus Unedo*. The southern sandstone chains are clothed with an extremely luxuriant and varied 'montebaxo,' varied near Cordova with forests of pines and cork oaks. The quartzose rocks of La Mancha are also covered by a 'montebaxo,' abounding in variety of forms, among the shrubs of which *Cistus populifolius* is distinguished. In Huelva Portuguese forms become associated with the other shrubs, for example, *Genista tridentata, Ulex genistoides*, with *Erica umbellata, Teucrium fruticosum, Helianthemum halimifolium*, &c.

The spring vegetation of the Sierra Morena has not yet been investigated, although it would be most interesting. The summer drought commences in July, and from that time till autumn few or no flowers are met with. Different kinds of bulbous plants, such as *Scilla maritima, autumnalis, Leucojum autumnale, Merendera Bulbocodium*, &c., flower in autumn, and seem very uniformly distributed.

The plateau of Madrid, occupying the centre of the peninsula, is separated from the preceding regions by the mountains of Toledo, which apparently very much resemble the Sierra Morena. On the north it is bounded by the long ridge of the Sierra de Guadarrama, rising to 7000 or 8000 feet, on which the snow lies during eight months of the year. The mean elevation of the plain is more than 2000 feet, and the mean temperature of Madrid is 59° Fahr., the mean of summer 77°, and of winter about 43°; but the thermometer always falls below the freezing-point in winter, so that there is skating almost every year on the ponds of the Retiro. It seldom falls below 22°, but in 1830 it was as low as 14°, and in 1802 $11\frac{3}{4}$°. In summer it occasionally rises, in still air, to 98° or even 105°. Rain falls only in winter and spring, with prevailing north winds, which are cooled by passing over the mountains; in spring, however, these winds alternate with the westerly and southerly, which characterize the summer, and are then accompanied by greater heat and drought. The autumn is warm throughout until December. The vegetation of herbaceous plants commences in the beginning of March, by the end the trees are in leaf and the cherry and lilac in blossom, and all herbaceous vegetation is over by the end of June, only certain shrubs withstanding the effects of the drought.

The plateau, which presents undulating ridges of low hills, is covered in the vicinity of Madrid mostly with wheat- and barley-fields, and being destitute of woods or even of shrubs, presents a most uniform aspect, and is everywhere bounded by the same limited horizon. The vegetable formations appear to be dependent on the nature of the soil, and thus fall into four classes, those of the clay, the gypsum, the sand, and the granite.

The argillaceous soil stretches from Madrid southward over the greater part of La Mancha, while the hills, from Aranjuez to Alcala, consist of saliferous gypsum, from which common salt effloresces. To the north and west of Madrid, as far as the mountains, the surface is composed of close-grained sand, without stones, which has the peculiarity of becoming consolidated by the drought, almost to the same degree as the clay. The granite soil occurs in the Sierra de Guadarrama, and blocks of it are scattered over the sand. Limestone is not met with in the neighbourhood of Madrid; it makes its first appearance in the east, towards Cuença, and with it the wide-spreading shrub-formations, or *montebaxo* of Catalonia, which are not found on the plain of New Castile.

The corn on the sandy soil is poor, on the clay it grows about 4 feet high; *garbanzos* (*Cicer arietinum*) and *garrobas* (here *Ervum monanthos*) are especially cultivated for food. The vine and olive only occur in sheltered places, the latter always small and poor. Meadows are altogether wanting; and even the pasture land at Manzanares is clothed solely with annual grasses and Leguminosæ, which are choked at the approach of summer by the thorny plants, such as *Centaurea Calcitrapa, Eryngium campestre, Ononis spinosa,* &c.; or where the ground is more marshy, are replaced by *Juncus acutus* and *Scirpus Holoschœnus.*

According to ancient chronicles, it would appear that the plateau of Madrid was formerly wooded, and remains of woods, in the form of stunted, widely scattered oaks, in particular *Q. Ilex,* are still visible on the sandy hills of Casa de Campo and of the Prado, mingled with leafless brooms (*Retama sphærocarpa, Sarothamnus scopa-*

rius); but these, with the willows, poplars, elms and ashes, growing on the banks of the rivers, together with some shrubs, such as *Tamarix gallica*, species of hawthorn, rose, bramble, buckthorn, &c., constitute the entire ligneous vegetation of the plain. That this want of wood is a consequence of the drought, is shown by the lofty growth of the plantations in the valley of the Tagus near Aranjuez, before alluded to, as well as of the more recently planted avenues in Madrid, which are moistened by irrigation.

The vegetable formation on the clay soil is somewhat varied. In the fields, the most abundant plants in early spring are *Brassica orientalis*, *Lathyrus erectus*, *Polygonum Bellardi*, &c., but these soon become choked by thistles and other thorny Compositæ, species of *Scolymus*, *Picnemon, Onopordum, Xanthium*, &c. Almost the only plant surviving at the end of summer is the cucurbitaceous plant *Ecballion*, which then ripens its fruit; *Crozophora* is also abundant. The uncultivated plains and hills are covered with low, shrubby, aromatic plants, to which the general term *tomillares* (from *tomillo*, thyme) is applied by the Spaniards; mixed with the plants of this class are other species peculiar to dry localities, such as *Minuartia*, species of *Astragalus, Echinops*, and the grasses *Cynosurus Lima* and *Stipa barbata*. By streams occur such plants as the mallows, *Althæa officinalis* and *Lavatera triloba*, with a few Cruciferæ, Sileneæ and Cichoraceæ.

The vegetation peculiar to the gypsum extends, with this formation, through the whole of Arragon. The steep declivities are clothed with patches of *Frankenia thymifolia*, with *Peganum*, Cruciferæ, Sileneæ, *Helianthemum squamatum*, and *Salsola vermiculata*. *Vella*

Pseudo-Cistus, Iberis subvelutina, Herniaria fruticosa, Centaurea hyssopifolia, and *Statice dichotoma,* are also characteristic plants. The ridges of the downs, as far as Aranjuez, are covered with the Esparto (*Macrochloa tenuissima*), used here, as in Valentia and the south, for a variety of purposes; with it are associated the rosemary, fritillaries, *Pimpinella dichotoma,* &c. Many isolated thickets appear, formed of *Quercus coccifera,* with *Rhamnus lycioides, Retama sphærocarpa,* and *Bupleurum frutescens.*

The sandy soil is characterized by the abundance and variety of Cruciferous plants, which in spring give a yellow colour to large tracts of cultivated plain. With this prevailing colour are mingled the blue of Boraginaceæ and the white of the camomile tribe. The most remarkable genera are: *Diplotaxis, Sisymbrium, Brassica* and *Sinapis; Anchusa* and *Echium; Anthemis,* and also *Linaria,* together with *Hypecoum, Rœmeria, Cerastium,* &c. As this rich covering of fresh vegetation disappears, the fields become overgrown with the tansy (*Tanacetum microphyllum*). The *tomillares* occupy extensive tracts, consisting of thymes, lavenders, *Santolina,* &c. In the spring, a multifarious growth of annual herbs and grasses is met with among these, and after its disappearance larger herbaceous plants, especially Umbelliferæ, spring up.

On the granite of the Sierra de Guadarrama, these *tomillares* ascend to about 4000 feet, the constituent plants becoming gradually changed. The increased moisture allows of the presence of many Central-European plants. Extensive pasture grounds for horned cattle, which are protected from the invasions of the flocks of sheep by fences (termed *dehesas*), are clothed with bushes of *Quercus Toza* and *faginea, Daphne Gnidium* and *Juniperus Oxycedrus*; the *Cisti* first appear here. On the whole, this part does not differ greatly from the plain. The higher mountain region, from 4000 feet to the summits, may be called the region of the brooms, since it is almost covered by *Genista purgans.* Solitary shrubs of juniper and *Adenocarpus hispanicus* occur here, upon the latter of which lives the true Cantharis or Spanish blister-fly. Some of the higher points rise above this broom region, and are covered by a thick, fine turf of *Festuca curvifolia,* mixed with the thrift, *Armeria juniperifolia.* Of alpine plants there are but few traces, such as *Saxifraga nervosa* and *hypnoides, Ledum hirsutum* and *brevifolium,* while the annuals of the sandy plain of Madrid flourish up here. About the mountain rivulets the turf is composed of *Nardus stricta,* with *Pedicularis sylvatica, Jasione carpetana,* and *Veronica serpyllifolia.*

On the northern declivity alone of this Sierra are pine forests met with, composed of the Scotch fir, and large tracts are covered with the brake fern. The Sierra de Gredos, the most westerly and highest elevation of this ridge, differs but little in its vegetation, and is still poorer in plants and more uniform in aspect.

We have not any detailed account of the vegetation of Old Castile, which, however, so far as the catalogues of plants go, appears to contain many more of the common North European plants. It probably forms an intermediate link between New Castile and the north coast.

A similar difficulty occurs in regard to the vegetation of Portugal, for although the general character of the country is tolerably well known, and most of the

x

indigenous species have, in all probability, been seen and determined, yet no botanist has systematically and accurately investigated the distribution of the plants in detail, or studied the peculiarities of the meteorological conditions minutely in the different regions. Our sketch of Portugal will thus necessarily be very imperfect.

The most southern part of Portugal is the province of Algarve, which is divided into three regions, running parallel with the sea-shores. The first is a line of sandy coast scarcely extending two leagues inland; this, which was originally a desert tract, has been converted by the industry of the inhabitants into a complete district of gardens, containing plantations of southern fruits, vineyards and corn-fields.

Groves of oranges and citrons lie between the ridges of the sand-dunes; woods of St. John's bread tree and olive, with wide-stretching, richly-leaved branches, alternate with plantations of figs and oranges, vineyards and orchards, and between these appear broad green meadows of cultivated fodder plants, surrounded by mulberry- and almond-trees; while here and there a proud palm waves its nodding crown of feathery leaves in the mild air, above this mass of evergreen vegetation.

The "glorious valley of Monchique," as it has been called, is said to be eminently beautiful, the luxuriant vegetation "refreshed by streams of the purest water, upon the banks of which the rhododendron grows profusely amid the lotus, the jonquil, and many varieties of *Scilla*, while the hills are covered with chestnuts of an immense size, and orange-trees bowed down by the weight of their golden fruit."

A pine forest of some extent occurs in one spot, in the

shade of which grows *Erica umbellata*; other characteristic plants are: *Empetrum album, Ulex Boivini* and
genistoides, Myagrum iberioides, Linaria præcox and
linogrisea, Scilla odorata, &c. The cactus-hedges are
entwined with *Aristolochia glauca*.

The hilly country or *barrocal*, extending up to 1000
feet, is much broken up, consisting of various conglo-
merates; it is fertile and well-watered; still a considerable portion of the good soil lies waste, and is covered
with *montebaxo*, with woods of the evergreen oak and
Quercus Ballota. *Erica lusitanica* and *australis* are
characteristic plants here.

The mountainous region is the terminal, undulating
prolongation of the Spanish Sierra de Morena, and like
that consists of grauwacke and clay-slate, excepting the
western portion, the Sierra de Monchique, which is composed of granite and basalt. Here the influence of the
geological subsoil is shown by the prevalence of the
shrubs of the Sierra de Morena on the sandstones and
clay-slate. Over the granite and limestone tracts, which
are limited in extent, "the earth is clad in its richest
apparel," and, according to an enthusiastic traveller,
displays in spring a scene of exquisite beauty; "besides
the rosemary, juniper, myrtle, lavender, and a thousand
bulbous plants disclosing their endless beauties, the
heaths, *Erica umbellata* and *australis*, with their brilliant and deep red blossoms, and the various *Cisti*, some
yellow, some of a rosy tint, some white as snow, and
others streaked with purple, embroidered the plain with
their variegated and delightful hues." But where the
sandstones and slates prevail, "the traveller may pass
through mountain defiles and over plains covered as far
as the eye can reach with the tall and unvarying *Cistus*

ladaniferus; and yet the graceful form of this plant, its green glistening leaves, its large, white, sleepy-looking flowers heavily spotted with purple, meeting the sight in every direction, are not without their influence on the mind." The monotony of the *Cistus* plains is sometimes varied by tracts covered with the dwarf palm, so frequent in the south of Spain, which, "like a dwarf aping a giant, is in some respects a caricature of the great eastern palm, yet with its elongated leaves and short, wild-looking stem, its appearance is picturesque enough."

The valleys of the Sierra de Monchique are wooded with chestnut-trees and cork oaks, among which grows the eastern *Rhododendron ponticum*. This mountain range does not rise higher than about 3800 feet, but the upper vegetation is subalpine, corresponding with the altitudes of 5000–6000 feet in Andalusia, which is probably a result of the cessation of the influence of the plateau here. The vicinity of the sea depresses the temperature in a regular vertical direction in normal cases; the high limits of vegetation in Andalusia, as over the interior of Spain, are exceptions, where the disproportionate height is accounted for by the heating influence of extensive tracts of elevated table-land.

A very different scene is presented on the north side of the Sierra de Monchique. There the wide, arid plains of Alentejo stretch out, a tract of country which from various causes is wild and little cultivated. In it dreary wastes of sand and heath alternate with low, hilly tracts, often covered with the wild olive and the *Quercus cocci-fera*, intermingled with cork-trees almost as huge as forest oaks. The *Cistus populifolius* is now and then met with, but the *C. ladaniferus* is the prevailing plant, covering extensive areas, and attaining a height of many feet.

Lisbon, with the hills of Cintra, have a world-wide reputation for beauty and luxuriance of vegetation, but unfortunately we have no accurate botanical description of the surrounding country. The favouring climate, where the cold of winter is scarcely felt, and the influence of the Atlantic shows itself in all its power in equalizing and tempering the heat, allows a wonderful variety of plants to flourish in the gardens, especially where the drought of the high summer season is guarded against by irrigation. The villages are surrounded by orange groves, cork woods, the roads lined with poplars, bays, and enormous willows, or overhung with oak and elm interspersed with box, forming a delightful shade. In the gardens of Cintra the open borders are filled with lemons, oranges, and southern fruits, while woods of enormous chestnuts abound, and on the neighbouring rocks the ferns of Madeira and the pelargoniums of Africa have established themselves and run into wild luxuriance.

The vineyards about Colares are often small and divided by stone walls, so as to break up the country into parterres as it were, like a large garden ; and amid these nature displays the greatest richness of vegetation. The olive, arbutus, the plane, and gigantic stone pines abound ; cork-trees everywhere environ the roads in the most fantastic shapes, with ferns growing on their huge trunks, and mistletoe hanging profusely from their branches. Jasmines load the air with perfume ; and various kinds of creepers, those especial marks of luxuriance of vegetable life, overrun the trees, sometimes covering them, oppressing them with their rank growth, and almost concealing their foliage.   Orange and lemon groves are mixed with Indian corn and water-melons, fruit-trees of every kind meet the eye in every quarter, and the vine, trained over trellis-work, instead of being topped and staked as in Germany, loses its stiff and formal aspect, and presents a graceful and elegant appearance.

The valley of the Tagus, extending in an irregular course from the hilly region of Estremadura to Lisbon, presents alluvial deposits at various points on both banks, its course having evidently varied at different epochs; but its encroachments have been most restrained on the north side by the range of hills which run from the Serra da Estrella to the Rock of Lisbon.   On the south side the alluvial plains are extensive, producing large quantities of maize and wheat.   In autumn, after the removal of the harvest, the open country is covered with a rank vegetation, consisting chiefly of *Xanthium spinosum, Datura Stramonium* and *Metel, Amaranthus albus* and *retroflexus, Sida Abutilon, Chenopodium olidum,* &c., and a large number of dried-up Cynaraceæ and Asperifoliæ, which apparently form the mass of the summer vegetation.

In waste places *Pistacia Lentiscus* and *Phillyrea latifolia* abound, supporting *Smilax mauritanica* and *Lonicera Periclymenum.*   About Santarem, as at Lisbon and elsewhere along the valley of the Tagus, the hedges are chiefly formed of the common *Agave americana,* planted about three feet apart, and as the older plants die after flowering, the young ones are placed in the intervals between them.   They are in full fruit in October.

To the north of the Tagus, above Abrantes, a district of totally different character is entered, presenting a gradual ascent among barren, rugged hills.   The orange and the agave are no longer seen, and the olive is confined to favourable situations; cultivation becomes less frequent, and the vegetation soon assumes the marked features which prevail without much change to the borders of Castile. With the exception of very limited tracts about villages, which are maintained in constant cultivation, the whole of the hilly district is covered with a brushwood composed almost exclusively of Cistaceæ, Ericeæ, Labiatæ, and two or three species of fruticose Leguminosæ. Of these, *Cistus ladaniferus* is almost universal, and forms often the sole vegetation of large tracts; *C. populifolius* is local; *Helianthemum halimifolium*, very common; *Erica scoparia, arborea* and *umbellata*, all common; *E. ciliaris*, apparently rare; *Calluna vulgaris*, uncommon. Of the Labiatæ, *Lavandula Stæchas*, or an allied species, is most common, with several species of *Thymus* and *Micromeria*. Of the Leguminosæ, the most interesting is *Genista tridentata*, forming a dense, dark green, tortuous mass; this, though very abundant in Portugal, seems to disappear soon after passing the frontier of Spain.

Ulex nanus occurs near Abrantes, but is soon lost. The common myrtle is occasionally seen, but much less frequently than in similar situations in Italy. The only tree which is at all common through these hills is the cork oak, which supplies an important article of trade throughout this part of Portugal. In some places *Pinus Pinea* forms groves of no great extent. The comparative scarcity of trees is explained by the practice which appears universal throughout the almost uninhabited districts of the interior of Portugal and Spanish Estremadura, of burning the dry surface of the country at the latter end of summer, chiefly for the purpose of obtaining herbage for the large flocks of sheep and goats which are scattered at wide intervals over the hills. Travellers coming upon such burnt-up tracts are be-smeared by the charred stems of the *Cisti* and other shrubs. As this process is repeated every five or six years, the effect is to reduce the perennial vegetation to an uniform period of growth, and to add very much to the monotony of the scenery. Although few traces of recent cultivation occur here, except in the neighbour-hood of the villages, there is evidence of the plough having passed over the greater part of these *dehesas*, at some perhaps remote period; and large tracts of *Cistus* and *Lavandula* are frequently seen growing in parallel lines which mark the furrows of the plough, looking as if the country were converted into a nursery garden for the growth of these plants.

The northern part of Portugal has been very little visited by botanists, and no sufficient materials exist for the description of it. The Douro, it is said, forms in some degree a boundary of vegetation, since the flora to the north of this, extending through Gallicia, is essen-tially distinct from that of the district of Lisbon. The great wine region of Portugal lies upon the banks of the Douro, north of Oporto, being situated in the provinces of Entre Minho and Trazos Montes.

Sect. 9.

Italy.

.........It is a goodly sight to see
What Heaven hath done for this delicious land!
What fruits of fragrance blush on every tree!
What goodly prospects o'er the hills expand!

THE peninsula of Italy forms a naturally defined region, bounded by the Alps on the north, and the Mediterranean and the Adriatic on the other sides; it lies between the 38° and the 46½° N.L., and is of elongated form, with a deep bay at the south-eastern extremity; the greatest length is from N.W. to S.E., and amounts to 130 geographical miles.

The Apennines stretch over the greater part of Italy as a long range of mountains, which may be regarded as a branch of the Alps. Their direction is at first W.S.W. and E.N.E., from the south-western portion of the Alps to Genoa, then W.N.W. and E.S.E. between Genoa and Bologna; here the direction again changes, and for the greater part of their extent, as far as Nicastro in Calabria, they are nearly parallel with the direction of the peninsula, that is, N.W. and S.E.; finally, they turn N.N.E. and S.S.W. in the south-western promontory next Sicily. The length of the Apennines is about 150–160 geographical miles, their breadth from 5–15. They have few lateral branches of importance; the largest are the spur in the north of the city of Aquila, and that which runs east of Sulmona to the Adriatic. But beside the Apennines, more or less connected with them, lie several masses, such as the Apuan Alps between Carrara and Fivizzano, and the range between the Pontine Marshes and the river Garigliano. Quite detached from the Apennines occur—Gargano on the Adriatic, Vesuvius, the Alban Mountains near Rome, and the Euganian Hills near Padua; the last three are of volcanic origin.

The mean height of the Apennines amounts to :—

	Feet.
From the Alps to Monte Cimone	3700
From Monte Cimone to Monte Sibilla	4700
From Monte Sibilla to Monte Matese......	6400
From Monte Matese to Monte Pollino	2600
From Monte Pollino to Nicastro..........	4800
From Nicastro to Aspromonte............	3200

Thus the middle portion, between Sibilla and Matese (the Abruzzi and a portion of the Papal territory), is the highest; here also occur the loftiest summits, and the range is broadest in this part.

The highest peaks in the Apennines are :—

	Feet.
Monte Cimone.................	7060
Alpe di Camporaghena	6537
Monte Sibilla	7188
Gransasso d' Italia	9493
La Majella	9314
Monte Pollino	7441
La Sila (approximative)	5300
Aspromonte	6375

The following is a list of the highest points lying about, or quite detached from the Apennines, and in the smaller Italian islands :—

	Feet.
Pizzo d' Uccello (Apuan Alps) ..	6131
Mont Amiata (near Sienna)	5825
Schiena d'Asino	4831

	Feet.
Monte Albano	3151
Gargano	3187
Vesuvius	4009
The Euganian Hills...........	1944
Elba......................	3290
Stromboli	2164

Between the Alps and the northern Apennines lies a very extensive plain, forming the Valley of the Po; the other larger level tracts are, that of Puglia on the Adriatic, the Roman Campagna, the plains of Pisa, and those near Naples.

The streams of the northern Apennines, the principal direction of which is W. and E., flow into the Po; the southern, maritime side has but inconsiderable rivers. The largest which arise in the Apennines lie more to the south; the Arno, Tiber, Garigliano and Volturno, on the west; the Pescara and Sangro on the east. But none of these rivers of the Apennines are of great size, and they are often dried up in summer from the want of rain or melting snow, thus essentially differing from the well-supplied streams of the Alps. The largest waterfall in the Apennines is that of Terni, 300 feet.

Besides the large lakes at the foot of the Alps, there are many among and at the foot of the Apennines, as the Lago di Bientina, L. di Perugia, L. di Bolsena, and the L. di Fucino; the last lies 2000 feet above the level of the sea. There are also extensive marshes, such as the Pontine, the coast region north of Pisa, near Comacchio, &c.

The principal mass of the Apennines is composed of a solid, grey limestone, but granite and other rocks occur both in the north and south parts. The detached mountains are mostly volcanic, among which Vesuvius and Stromboli are still active.

Sicily (between 38°–36½° N.L.) consists in great part of an undulating mass of mountains, of about 1500–2000 feet in height, higher ridges rising above this. Among these, the largest and highest are the Madonians, attaining 6500 feet, and composed chiefly of the Apennine limestone. The isolated cone of Etna rises far above these, its summit reaching 11,300 feet, a point exceeding the loftiest in the Apennines.

Sardinia (39°–41°) has a flat, elongated range of mountains on the east side, the highest peak, Genargentu, rising to 5984 feet. In the west lie extensive plains, with a few detached elevated masses.

Corsica (41½°–43°) is very mountainous, and the points are of great height; Monte Rotondo attains 9037 feet, Monte d' Oro 8676 feet. Both Corsica and Sardinia are chiefly composed of primitive formations. The temperatures may be drawn from the following summary :—

	Lat.	Ann.	Winter.	Summer.
Milan	45½	55	36	73
Bologna	44½	57	36	76
Florence	43½	59	44	75
Nice	43½	60	49	72½
Rome	42	60	47	73
Naples	41	63	50	75
Palermo	38	64	52	75

The north and south sides of the northern Apennines afford a striking contrast; Lombardy and even Bologna have a comparatively cold winter and warm summer (a continental climate). From Bologna to Florence the winter temperature rises about 8°, from Bologna to Nice

13°, while the summer heat remains the same, or even falls lower. This difference in the distribution of heat through the year exercises an essential influence both upon the wild and the cultivated vegetation, very striking to the traveller journeying from Piedmont towards Genoa or Nice, from Modena to Lucca, or from Bologna to Florence. The climate and vegetation characteristic of the Mediterranean coasts are not met with until the northern Apennines are passed. But from Florence the warmth of the winter increases considerably towards the south, the annual mean little, and the summer heat not at all. It should be noticed, however, that the temperature of autumn rises more considerably towards the south than that of summer, or, in other words, the summer heat is prolonged into the autumn. Thus, in Milan and Bologna the month of June is warmer than September, in Florence, in Rome and Naples they are equal, while in Palermo and Cagliari September is hotter than June. The annual mean of Palermo equals the summer heat of Copenhagen, the winter average equals the May temperature of the latter city; the spring of Palermo is almost warmer than that of September in Copenhagen, or of the summer in Edinburgh; and the autumn temperature of Palermo higher than that of the warmest month in Copenhagen. And yet the climate of Palermo is cooler than that of other parts of the coast of Sicily, because that city is surrounded by mountains on the south and exposed to the north winds. The mean temperature of the south coast probably reaches 68°. Frost is rare in Palermo, and during forty years' observations made there the thermometer did not fall below the freezing-point.

At the foot of the principal crater of Etna there is a house (called the English House, or the House of Ge-

mellaro), probably the highest habitation in all Europe.
This remains buried in snow even in June, and the fresh
snow falls again in August. According to observations
made there in July and August, the mean temperature
of these months was only $43\frac{1}{4}°$, while simultaneous observations at Catania, on the sea-shore, gave $80\frac{3}{8}°$.

The quantity of rain is very great at the immediate
foot of the Alps, 50 or 60 inches; but it diminishes so
considerably towards the south of the plain of Lombardy,
that it scarcely amounts to half that at the northern foot
of the Apennines. On the immediate slope of these
again it is very large, and then diminishes gradually to
Palermo, where it only amounts to 21 inches. Under
otherwise equal circumstances, the west side of the Apennines receives more rain than the east. As in Spain,
the summer-rains of Italy are inconsiderable, except in
the plains of Lombardy, and they decrease in their proportion to the annual fall towards the south. The proper
rainy season here is the latter part of autumn and the
winter. The following table gives a view of these proportions, the figures giving the per-centage to the annual
quantity of rain :—

	Lat.	Summer.	Autumn.	Winter.
Padua	$45\frac{1}{2}$	0·26	0·31	0·21
Florence	$43\frac{1}{2}$	0·14	0·35	0·28
Pisa........	$43\frac{1}{2}$	0·12	0·39	0·28
Rome	42	0·11	0·35	0·30
Naples......	41	0·11	0·39	0·28
Palermo	38	0·06	0·35	0·37

Snow is rare in the plains south of Naples, and melts
as soon as it falls. On the higher mountains of the
Abruzzi, on Etna, and on the Corsican mountains, the
snow lies almost all the year, and even here and there

little masses never melt; thus the highest peaks only approximate to the line of eternal snow, and no glaciers are formed.

The *sirocco* is a hot parching wind coming from Africa, and rendering the atmosphere thick. The *malaria* prevails in the warmest seasons in many places; these are either uninhabited, or the people leave them during this part of the year. These regions are partly morasses, in which the decaying organic matters infect the vapours rising from the stagnant water, as in the Pontine Marshes; partly also large and for the most part dry plains, like Puglia, the Roman Campagna, and the western plains of Sardinia. The rice-fields, which are flooded with water, also render the atmosphere unhealthy.

To Italy has the glance of the European botanist been ever directed, as the region where the more modest forms of vegetable life which mostly clothe our continent, become mixed with some of those glorious productions which characterize the tropics; and where he may in some degree realize a conception of that luxuriance of vegetation which flourishes in all its splendour only in other quarters of the globe. The Spanish peninsula is indeed a worthy rival of Italy, especially in certain districts, but has been, until a recent period, comparatively unknown to botanists; while the Isles of Greece must yield the palm of wealth of vegetable life to Southern Italy, where the climate is so much more highly favoured by local circumstances, modifying the influence of mere latitude.

Greece and Spain represent only by a couple of species of *Mesembryanthemum* the succulent plants of the burning plains of Africa, and the Balanophoreæ only by *Cy-*

tinus hypocistis; while the Italian islands possess the curious *Stapelia europæa*, a type of the strangest vegetation of Southern Africa, and in *Cynomorium coccineum* afford an important link connecting our parasites with the gigantic *Rafflesia* of the East Indian Islands.

Lombardy presents a tract of the greatest possible interest in reference to geographical botany, from the variety of conditions that prevail there ; while in the kingdom of Naples, particularly the southern part, and still better in Sicily, we meet with conditions approaching more nearly to those of the equatorial countries than in any other part of Europe, with perhaps the exception of one or two points on the coast of Spain. Space being limited, we shall confine our survey of Italy to these portions of its surface.

The limits of the Lombardic region, of which we shall first treat, extend from the river Sesia, from west to east, to the Adige ; on the north they run along the watershed of the Alps, forming a vast curve from Monte Rosa along the ridge between the above-named rivers, to Mont Adamo (including the canton of Tessin and a part of the Valesian Tyrol). On the south the Po forms the boundary. The country belongs to the region of that flora of Upper Italy which has been termed the zone of transition between the Alpine flora and that of the Mediterranean, and within its limits are found remarkable boundary lines between these two floras, which would be stil! more strongly marked had not the hand of man interfered so much and for so long a time in effacing the primitive conditions.

The combined influence of the sea surrounding Italy on three sides, the periodic winds, and the yearly and daily changes of temperature corresponding to its latitude, cause the development of the plants of the Mediterranean flora in all parts of the peninsula; but this southern flora cannot be regarded as predominant in Upper Italy. The great valley of the Po constitutes a sort of funnel, the point of which rests against the Cottian Alps, and which is enclosed on all sides by high mountains, excepting at its broad base on the east, which spreads out to the sea; from thence the hot, moist south-east winds flow freely in, till, penetrating far westward and meeting the icy currents which pour down from the vast snow-fields in the N.W., they gradually lose their mild influence, everywhere evident in the Venetian territory, but almost lost beyond the river Sesia; so that the flora there only retains its southern character in the vicinity of the Po, in the defiles of the adjacent Apennines, and opposite to them. The Valley of Aosta forms a remarkable exception.

In agreement with this configuration of the country, the line of demarcation between the Alpine and Mediterranean floras follows an oblique and serpentine course from S.W. to N.E. In the neighbourhood of Turin it approaches the Po, and stretches thence S.W.–N.E. to the vicinity of Brescia. Here it coincides with the branches of the Rhætic Alps, which extend southward, converging towards it, and are immediately exposed to the influence of the maritime winds; running eastwards on the lower portion of them, it curves suddenly to the north near Selvapiana di Prandaglio, on the banks of the Chiesi, and may then be traced to a considerable height on the mountain ridges as far as Tyrol, behind Salo and Garguano, where the Mediterranean flora is developed under a true tropical temperature, only interrupted in winter by the storms of cold wind pouring down from Baldo.

Y

The course of vegetation in Lombardy therefore deviates considerably from that of other continental tracts of central Europe. A remarkable phænomenon is also afforded by the exceptional regions around the many lakes of Upper Italy, which lie within this region.

The influence of the sea in the development of certain forms of vegetation in the countries lying around the Mediterranean depends in part upon the peculiar constituents (salts, &c.) with which the soil was in former ages and in part is still impregnated. But a much greater influence is exerted by the equable distribution of heat, caused by the vicinity of a large expanse of water, and this proposition is strongly confirmed in Lombardy around the lakes and over the tracts of rice-fields. In tracing the course of vegetation from the most western basin, the Lago d' Orta, to the most eastern, the Lago di Garda, it is readily perceived that the characters of the vegetation of the tracts surrounding them are subject to the influence of two laws,—one the general rule already announced, that the forms acquire a more southern aspect towards the east; the other a special one, compounded of three factors, namely, the relative height above the level of the sea, the exposure to the two opposite currents of air, and the position of the surrounding mountains as reflecting surfaces for the rays of heat; for these numerous lakes differ in all these respects from each other. It will suffice to mention one or two instances. Around the Lago d' Orta we find no trace of a Mediterranean flora—no olives; on Lago Maggiore at present no longer any olives, but a most luxuriant garden flora in the low Borromean islands, *Agave americana* growing and flowering in the open air. On Como the olive ascends 1600 feet above the sea, the vine nearly twice as high. On

Lago di Garda, the line of vegetation of the orange (which is protected in winter from the icy N. and E. winds from the Tyrol and Baldo) rises to some 1200 feet above the sea in the closed-up valley of Bogliaco, and the line of the olive rises there to some 2000 feet; in the neighbourhood of Maderno the whole of the sloping side of a valley may be seen densely clothed with naturalized agaves. These lake districts evidently constitute isolated and exceptional regions of vegetation, betraying the vicinity of the Mediterranean flora, in the midst of the Alpine territory.

All these isolated districts may indeed have been united in former times by a narrow line of olive-grown mountain ridge, stretching along the extreme southern declivity of the Lombard highlands; isolated remnants in the neighbourhood of Pusiano, near Montevegghia and around Brescia, seem to afford evidence of this. But the region where the Mediterranean forms of vegetation gain the upper hand must now be sought in those districts where we meet with the submerged tracts of the rice-fields, often extending farther than the eye can reach; although it must be observed at the same time that between the Sesia and the Adda, the predominance of southern forms is quite confined to the rice-fields, and the strips of dry land immediately adjacent or surrounded by them. When it is remembered that the culture of rice occupies, in Lombardy proper alone, a great many square miles of surface, which consequently exhibits an almost uninterrupted sheet of water from the commencement of spring until autumn, the influence of the cultivation of rice on the configuration of the physiognomy of vegetation in Lombardy will no longer appear surprising. The rice region is almost everywhere adjacent to the Po, and on the other side of that river the vegetation is in all cases distinctly southern.

M. Cesati has given us a lively picture of the gradual transition from the alpine vegetation to that of the Mediterranean plains, and we shall draw freely upon this for our illustration. He supposes a traveller to descend from the Alps, where the gates of Italy are surrounded by snow and ice in the midst of summer, into the fertile plains of Lombardy. The course may be either over the glaciers of Tyrol, through the valleys of Camonica, Rabbi or Sacra, to the lakes of Iseo, Idreo or Garda, or at the opposite end of the region over the grim deserts of the Simplon or St. Gothard to Lago Maggiore ; or, avoiding the fearful passes of the Via Mala, may pass by the Splügen, or over the inhospitable Stilfser Joch into the valley of the Adda, to the delightful banks of Lake Como ; in any case, the botanist will find almost the same general course of vegetation.

On the highest passes, on the borders of the glaciers, along the steep bordering cliffs, on the last patches of alpine pasture, many plants of the northern slopes accompany the traveller for some distance down the south side, such as *Carex bicolor, irrigua, curvula* and *atrata, Luzula lutea* and *spadicea, Juncus Jacquini* and *trifidus, Lloydia serotina, Salix herbacea, helvetica, reticulata* and *retusa, Linnæa borealis,* many species of *Artemisia, Gnaphalium, Achillæa, Hieracium, Pedicularis, Androsace* and *Primula,* several Potentillas and Gentianaceous plants, with *Phaca, Sibbaldia,* &c.; but the northern botanist misses *Elyna spicata, Carex capitata, chordorhiza, grypos* and *tetanica, Trientalis, Juncus stygius, Campanula pulla* and *Zoysii, Wulfenia, Swertia carinthiaca, Silene pumilio, Ledum,* &c.; while allied genera and species which are wanting, or extremely rare, on the German side become abundant, such as *Carex baldensis, alpestris,* &c., *Valeriana tuberosa,* fresh species of *Artemisia, Hieracium lanatum,* numerous Campanulas, such as *Raineri, Elatine, elatinoides, cenisia, alpina,* &c., *Veronica Allionii, Viola Comollia* and *nummularioides, Ononis cenisia,* &c. *Braya* and the northern *Arabis* and *Draba* are replaced on the Italian side by *Iberis saxatilis, Barbarea bracteosa, Cardamine asarifolia* and *Dentaria heptaphylla.*

In the lower region of the Central Alps (the mountainous region), as in the similarly situated fore-Alps, a difference is perceived in the vegetation, which, however, cannot be at once marked by a dividing line ; for the allied forms create a deception which at first leads one to suppose that the well-known northern plants still continue to accompany the observer. Shortly before, *Cytisus alpinus* may have been gathered on the southern slope, but now it seems to have acquired a strange aspect that might be hastily attributed to a difference of soil and climate ; but if we approach to gather the golden blossoms, the scent tells us that we have found *Cytisus Laburnum.* By degrees, *Cytisus radiatus, glabrescens, purpureus* and *sessilifolius,* with *Medicago carstiensis,* usurp the place of the leguminous plants which flourish on the Swiss and German sides.

Primula Candolleana is exchanged for *P. glaucescens, Saxifraga Burseriana* for *S. Vandellii, Silene acaulis* for *S. Saxifraga.* In place of *Potentilla norvegica* appears *P. graminopetala* ; *Allium pedemontanum, Laserpitium nitidum, Gaudinii* and *peucedanoides, Molospermum cicutarium, Pæonia, Euphorbia variabilis* and *Viola heterophylla* are new acquisitions. The delighted collector seeks for *Zahlbrucknera paradoxa* and finds the delicate *Saxifraga arachnoidea*, which penetrates through the Val di Ledro from Tyrol as far as the mountains of Val Trompia, while on the hills of Brianza (between Adda and Lambro) and Verona, it yields again to its near relative *S. Ponæ.* Instead of *Pæderota lutea, Primula integrifolia, Anemone sylvestris* and *Mœhringia polygonoides,* appear *Pæderota Bonarota, Primula Polliniana, Anemone baldensis* and *Mœhringia Ponæ.*

The fine grass, *Festuca spadicea,* here finds a worthy representative in *F. spectabilis. Ranunculus Villarsii* and *Thora, Astrantia minor, Saxifraga Cotyledon* and *exarata,* are not new, and for a time may mask the transformation, only their more frequent appearance and more luxuriant growth will be perceived.

Soon the traveller advances into the rich woods of broad-leaved chestnuts, under whose hospitable shades *Erica arborea, Scabiosa gramuntia, Centaurea nervosa* and *austriaca, Carex alpestris,* &c. predominate, while *Cratægus torminalis* and *Azarolus, Rhus Cotinus,* and holly form the underwood. Lower down the mighty *Celtis* lifts its crown, and in the eastern chains the evergreen oak and *Quercus Esculus,* beneath which the odoriferous *Syringa* (*Philadelphus*) displays its snowy blossoms. At a winding of the valley the view of an enchanting lake opens, and while the icy crest of the giant Alps, round which stormy winds are brawling, still hangs as it were over the head of the wanderer, his cheek is fanned by the mild air of the olive groves, playing amidst the mournful cypress, the classic laurel and the slender stone-pine. A step further, and he no longer rests beneath the scanty shade of the pinaster, but seeks the shelter of the rich variegated foliage of the sycamore. Among the long lines of vines and mulberries of the well-cultivated slopes, he beholds on every spot yet unreclaimed by man, the poetic myrtle, the pomegranate with its glancing leaf, its flaming blossom and dappled juicy fruit, the Italian beech, the white-blossomed ash, and the evergreen box. Then *Pistacia Terebinthus, Cercis, Phillyrea,* and *Prunus Malaheb,* with *Rhus Cotinus* in abundance. And among the bushes, in the hollows of the cliffs, on the pebbly margin of the torrent, amid the wildly luxuriant grass which clothes the unconquerable portions of the soil, springs up in all directions the herbaceous flora of the south : *Campanula spicata, Cytisus argenteus, Ononis Natrix, Silene insubrica* and *italica, Althæa cannabina* and *narbonensis, Satureia montana* and *hortensis,* &c. &c. On the banks, in the hedges, in the chinks of the stone walls which bear up the fruitful soil in terraces, upon the hill sides, from the cracks of ruined buildings, shine the red berries of the *Ruscus* beside the white-spotted leaves of the *Arum,* and around them flourish *Andropogon Gryllus, Verbasca,* Silenes, Linarias, and numerous southern umbellate plants.

At length the zone of olive and citron is attained, and these clothe the land even to the sea-shore with a gloomy veil of foliage. It would indeed be vain to seek here for the rare plants which luxuriate beneath the tall olives of the Ligurian coast, such as *Tolpis, Andryala, Lavatera punctata* and *olbia, Hippocrepis multisiliquosa,* &c., but the botanist will be rewarded by the southern lichens which abound upon the hard Jura limestone.

But these lovely landscapes which afford a charming foretaste of the southern flora, are but so many oases on the border of the Lombardic plains, the fertility of which is in truth proverbial, but which are uniform enough to drive the botanist to despair. Yet in the wooded hollows of the higher parts of the plain, and even in the fields, he may gather many interesting plants, such as *Lathyrus angulatus* and *hexahedrus*, many trefoils and grasses, *Ranunculus velutinus* and *insubricus*, &c. Then follow the irrigated meadows which bear witness of the enduring industry of the Lombards, and yield six crops of hay each year. There the botanist may pick *Sorghum halepense*, *Eragrostis pilosa*, *Alopecurus utriculatus* and *Cyperus longus*, and in dissatisfaction pursue his course toward the wide-spread rice-fields which accompany him as far as the Po. In these vast expanses of water, which in spring, when the delicate grain sends up its tender germs, are covered with a lovely light green carpet, live a host of social plants, such as species of *Cyperus*, Sedges, Potamogetons, *Xanthum italicum*, with the strange Marsileas and Salvinias.

The line of demarcation of the two floras is well-marked in the hilly tracts. The vicinity of Brescia is the true turning-point, where the vegetation of the valley of the Po completely changes its aspect. Through the defiles of the Adige the Mediterranean flora penetrates into the heart of the Valesian Tyrol, as the presence of *Bonjeania hirsuta*, *Hypericum Coris*, *Ptychotis heterophylla*, &c. testifies. Nearer to the Adriatic, the projecting side ridges of Baldo and the Beric and Euganean Hills, where *Arbutus Unedo* displays itself in full beauty, and even *Cistus laurifolius* surprises the botanist, the marsh flora is diverted completely to the south in the barren region of Tartaro and the Lower Po, where *Aldrovanda vesiculosa* occurs in its most extensive habitat.

Between the Sesia and the Mincio, however, that is, in the largest extent of the Lombardic region, the southern flora of the dry land of the plains is pushed further south through the cultivation of the land, than it would be in correspondence to the above-mentioned line of demarcation. On the Mincio, and eastwards of it in the low grounds of Tartaro, the number of marsh plants is increased by the following, partly rare, species : *Erianthus Ravennæ, Hibiscus roseus* and *Trionum, Arundo Donax, Dichostylis Michelianus, Stratiotes aloides, Chara ulvoides, Vallisneria spiralis* and *Aldrovanda vesiculosa* ; the two last occur in other stations. Finally, the Mediterranean vegetation appears on the banks of the Po with *Corispermum bracteatum, Anemone coronaria, Bupleurum protractum, Scirpus atropurpureus,* &c. Farther to the east marine plants are not wanting, such as glassworts and the like, with *Jasione Sicula, Aster Trifolium, Honckenya maritima, Atriplex triangularis,* &c.

There are no true mosses or moors such as are found in Northern Europe, therefore such plants as the cranberry, *Ledum palustre,* many willows, with *Malaxis, Lobelia Dortmanna* and *Wahlenbergia hederacea,* find no station here. The bog-moss is indeed indigenous, but occurs in isolated masses ; the half peaty low-grounds of the region afford *Rhyncospora alba* and *fusca, Schœnus nigricans,* northern sedges, *Herminium Monorchis, Gladiolus triphyllus,* three Droseras, cotton-grasses, &c. The great consumption of the peat, in a country so poor in fuel, will ere long limit or extirpate this exceptional flora. And thus the Lombardic flora gradually becomes fundamentally changed. In no portion has it suffered more than in the wood plants, since the wide tracts of forest have wholly vanished from the plains. The woods of Merlata and at Seveso, which scarcely half a century ago reached to the neighbourhood of Milan, still afford *Mœnchia*

erecta, Corydalis fabacea and *Trifolium squarrosum.*
But there will soon be no more traces of these, as on the
mountains, especially in the province of Como, most of
the deciduous forests and pine woods have disappeared.
The only scattered remains of the primitive forests are
to be found in Valesian Tyrol, near Campiglio, in Val
Taleggio (Bergamo), and in some few of the more distant
valleys of the Canton Tessin and the Valteline.

The flora of Lombardy contains 2568 indigenous
species, with 71 evidently naturalized from cultivation.
They include members of 129 natural families, exceeding
the German flora by 14, which is accounted for by the
absence, on the one hand, of the Lobeliaceæ and Myri-
caceæ, and the addition, on the other, of Capparideæ,
Zygophylleæ, Cæsalpineæ, Granateæ, Philadelpheæ,
Myrtaceæ, Cactaceæ, Ebenaceæ, Jasminaceæ, Phytolac-
caceæ, Laurineæ, Juglandeæ, Dioscoreæ and Bromeliaceæ, with Mimoseæ and Hippocastaneæ, probably introduced. The Swiss flora, numbering 116 families, is related to the Lombardic by the addition of the Cactaceæ,
Jasmineæ and Dioscoreæ, from its ranks, the German
Lobelias and Myrica being likewise wanting. The proportions of the principal families to the entire flora contrast strongly with those of Northern Europe : viz. Com-
positæ form $\frac{1}{8}$, Grasses $\frac{1}{13\frac{1}{2}}$, Leguminosæ $\frac{1}{14}$, Cruciferæ $\frac{1}{20}$,
Cyperaceæ $\frac{1}{22\frac{1}{2}}$, Umbelliferæ $\frac{1}{23}$, Caryophyllaceæ and Scrophulariaceæ each about $\frac{1}{24}$, Labiatæ $\frac{1}{26}$, Rosaceæ $\frac{1}{28}$, Ranunculaceæ $\frac{1}{29}$, Orchidaceæ $\frac{1}{47}$, Saxifrageæ $\frac{1}{73}$, Salicineæ
only $\frac{1}{83}$, and Ericaceæ only $\frac{1}{128}$.

The city of Naples, situated between the Apennines and
the sea, and consequently exposed to alternations of the

north and south winds, is subject to exceedingly rapid meteorological changes, rendering its climate very changeable and uncertain. In the winter, autumn and spring, the most variable seasons, it is not uncommon to see the thermometer fall and rise between 56°–59° and 41°–44° Fahr. during the day; and the sky alters with surprising rapidity from the greatest serenity to cloudiness, rain and storms.

Generally speaking the spring is very short, and often undistinguishable from the summer, from the great heat that is felt directly the winter has disappeared. Nevertheless, in some months of this season, ordinarily May, the changes of the atmosphere are so instantaneous, that in the same day the heat may be suffocating at one hour and a stove be required in the next. The winter is always rainy, but very mild. Cold is not felt until after Christmas, and January and February are the months in which the temperature sometimes, but for a few days only, becomes excessively cold. Vesuvius and the mountains visible from Naples are then covered with snow, which, however, melts in a few days. Snow very rarely falls in the city of Naples, and, when it does, never exceeds a few inches in thickness. It is more frequent on the mountains of the Terra de Labono and others which are visible from Naples, and remains on the ground some days. But it snows abundantly during the greater part of the winter on Mont St. Ange de Castellamere, whence ice is procured for the consumption of the capital.

The thermometer has been observed to descend as much as 11° below the freezing point in Naples. On the coasts of Calabria the thermometer never falls to the freezing point, while in the Abruzzi it will sometimes indicate 11° to 13° below the freezing point.

The westerly winds and the vicinity of the sea keep the air cool to a certain extent, and render the heats of summer less insupportable. In this season the thermometer is almost constantly between 77° and 86°, and only ascends as high as 88°, or even 95°, for a few days.

Much greater heat is experienced in Calabria, where the thermometer rises to 100°, and keeps between 86° and 95° for a long period.

Storms and hail are very frequent in all the provinces of the kingdom of Naples, probably on account of the influence of the many mountains on the electricity of the atmosphere.

The physical arrangement of the principal mountains, together with the geographical position of the peninsula, between the Adriatic and Tyrrhenian Seas, may explain the difference observed between the climates of the two opposite sides of the kingdom of Naples. It often happens, indeed, that with the prevalence of a south-west wind, Naples and the west coasts are deluged with rain for whole months, while not a drop falls in the eastern region ; and on the other hand, when the Greek wind is blowing there may be much rain, and even abundance of snow, in Le Poulle, while the weather is perfectly fine in Naples.

The Abruzzi, some of the summits of which almost rival the Alps in elevation, present well-marked zones of vegetation, extending from the sea-level to the cessation of vegetable life. In the other regions of Naples, where the mountains are less considerable, analogous conditions may be observed ; special differences, however, being produced by variation of exposure ; particularly seen in contrasting the eastern with the western sides, and the north and south extremities of the kingdom. M. Tenore divides the Neapolitan mountains into ten regions of vegetation, which he describes somewhat as follows :—

1. *Region of the maritime plains.*—This region is very little elevated above the sea, and is principally composed of bare tracts of sand, over which marine or volcanic products are frequently projected by the violence of the waves, which at the same time bar the passage of the torrents flowing from more elevated parts, and thus cause the production of pools and marshes, with which these plains abound. They are consequently almost entirely deserted and uncultivated, like the coasts of the kingdom, formerly so crowded with populous and flourishing cities. In those once enchanted spots where the public authorities in ancient times ordered the destruction of all cocks, lest their morning crowings should disturb the peaceful sleep of the happy inhabitants, the tired traveller cannot now close his eyelids without exposing himself to the risk of deadly fever. In vain are sought, among the pools and brushwood which now cover this pestilential soil, traces of the voluptuous Sybaris, the famous Heraclea, or the magnificent Metapontus. Of all those celebrated cities of antiquity we have now but the testimony of history. The vegetation of such a region could not be expected to prove very interesting. The principal trees met with are willows and poplars ; the shrubs are the lentisk, phillyrea, tamarisk, juniper, with *Ephedra distachya, Cistus villosus, Daphne Gnidium, Passerina hirsuta,* &c. The herbaceous plants are mostly of maritime character, some identical with those of our own shores, others belonging more particularly to the south. Of the former are the sea-holly, the sea-convolvulus, many Atriplices, with the southern *Echinophora spinosa, Cheiranthus tricuspidatus, Convolvulus Impe-*

rati, Salsola Tragus, &c., all growing in the sands, about
the sea-level. On the maritime rocks are found, besides
the usual European plants of such localities, the *Mesem-
bryanthema nodiflorum* and *crystallinum*, and *Aizoon
hispanicum*, &c. The brackish marshes abound with the
glass-worts, Atriplices, &c.; and on the borders of the
dikes and streams occur many southern grasses, such as
*Crypsis aculeata, Rottböllia spathacea, Agrostis fron-
dosa*, &c.

2. *Region of the Mediterranean plains.*—The soil of
this region, which is either sandy, calcareous or argillaceous, according to the predominance of these substances
in the subjacent rocks, rises gradually into hills, attaining an elevation of some 300 feet above the level of the
sea. Here the elm, the mulberry, maple, and wild pear
are the principal trees, while the fields afford shrubs of
the *Rhamnus Alaternus*, the Christ's thorn, blackthorn
and spindle-tree. The clefts of the rocks near the sea are
decked by the shrubby *Medicago arborea* and *Euphorbia
dendroides*. The most remarkable of the characteristic
herbaceous plants are the *Chenopodium ambrosioides,
Daucus mauritanicus*, species of *Centaurea* and *Carduus*,
with the common plants of Central Europe.

3. *Region of the hills.*—This region extends from
300 to 900 feet above the sea. The argillaceous, sandy
or tufaceous soil by which it is covered is subject to variations from the admixture of the constituents of the primitive secondary and volcanic rocks which roll down from
the neighbouring mountains. The porous lavas form a
portion of the soil of this botanical region, and when of
argillaceous nature, like that of the Vesuvian lavas, the
scoriaceous surface becomes decomposed in a few years
and acquires a dense coating of lichens, especially of

Stereocaulon vesuvianum and *Cetraria islandica.* These lichens, attacking the surface of lava, favour the gradual decomposition, which produces mould, filling up the cracks and inequalities of the lava. The first plants which establish themselves on this soil are the brake fern, *Spartium junceum* and *Salvia bicolor,* and by degrees the earth becomes capable of supporting every other species of the climate. In this region the trees of the south begin to assert their claims; the olive, the evergreen oak, and the stone pine, throw the wild pears and crabs into the shade, while the leguminous trees, *Cercis Siliquastrum* and the laburnum, add their brilliant blossoms to the decoration of the woodlands. Among the under-shrubs, the bladder senna, the brooms, *Spartium Scoparium* and *Genista candicans,* with *Salix capræa,* are most characteristic. The most striking herbs of the fields are : *Cynanche vincetoxicum, Globularia vulgaris, Carlina lanata, Sideritis syriaca, Salvia Sclarea,* &c. ; those of the mountains on the borders of this region, *Campanula fragilis, Rumex scutatus, Drypis spinosa,* &c.

4. *First region of woods.*—Extending from 900 to 2400 feet; this zone is almost entirely covered with trees, and especially those of the forest kinds, so that only a few shrubs or herbaceous plants grow in the less shaded places. The soil of this region is for the most part vegetable earth composed of the accumulations of decayed leaves of the forest-trees. These last are the chestnut and oak (*Q. Robur* and *Cerris*), the sycamore, pears and crabs, with the mountain ash. The shrubs are more decidedly southern, as *Cistus salvifolius* and *incanus, Mespilus domestica* and *pyracantha, Rhus Cotinus,* &c. Among the few characteristic herbs, *Cnicus Acarna* and *Silene Armeria* may be noted.

5. *The second region of woods,*—extends from 2400 to 3600 feet, and does not differ from the preceding in the composition of the soil, but is characterized by the appearance of the beech ; besides this tree, the Conifers, *Pinus Laricio* and *brutia,* the Scotch and silver firs, occur abundantly ; the shrubs are those of northern Europe, such as *Pyrus Aria, Vaccinium Myrtillus,* &c. ; while in the valleys, buried in the recesses of the wood, a rich herbaceous flora is met with, composed of such plants as *Pæonia officinalis, Delphinium fissum, Lilium Martagon, Atropa Belladonna, Aquilegia vulgaris* and *viscosa, Dentaria heptaphylla* and *bulbifera, Dianthus monspeliacus, Saxifraga rotundifolia,* &c.

6. *The mountainous region.*—This, which might also be called the region of meadows, from the carpet of verdure which everywhere clothes it, abounds in herbaceous plants, and is almost devoid of large trees. The turf which ornaments the soil is covered with snow in winter, and furnishes an abundance of nutritious food to the flocks in spring, when they are driven up from the lower regions. The rocks which form the slopes of this region, extending from 3600 to 4800 feet, are covered by a layer of vegetable mould only a few inches in thickness. The monotony of the surfaces is broken by a few trees of *Pinus Mughus* and *Juniperus Sabina* ; in the pastures, which are chiefly overgrown with species of *Agrostis* and *Festuca,* are observed *Statice Armeria, Globularia cordifolia, Astragalus montanus, Trifolium ochroleucum, Alchemilla alpina, Ranunculus brevifolius, Gentiana acaulis, Pedicularis rosea* and *foliosa, Campanula petræa* and *graminifolia, Astragalus areolatus, Hippocrepis glauca,* &c.

7. *Lower alpine region.*—This consists of the peaks

of the mountains rising above the preceding region from 4800 to 5400 feet: they are almost wholly composed of steeply escarped rocks, presenting only sparing traces of vegetation in the crevices, or in the deposits of soil accumulated in corners, or hollows of the rock. Among the plants peculiar to this region are: campanulas of the preceding, with *Astragalus alpinus, Viola montana, Linum campanulatum, Bunium petræum, Soldanella alpina, Valeriana saliunca, Sison flexuosum,* &c.

8. *Middle alpine region.*—This region, rising to 6000 feet above the sea, is, like the foregoing, devoid of trees, and only presents scattered thickets of wild shrubs and a few herbaceous species. The principal plants are: *Salix retusa, Arbutus Uva-ursi, Dryas octopetala, Gentiana verna, bavarica* and *acaulis, Sempervivum arachnoideum, Primula villosa, Erigeron alpinum, Arnica Bellidiastrum, Saxifraga glabella, cæsia, cotyledon, Aizoon, Iberis saxatilis, Alyssum tortuosum, Silene acaulis, Anemone alpina, narcissiflora, Polygonum viviparum, Adonis distorta,* &c.

9. *Upper alpine region.*—The elevation of this region extends to 7000 feet above the sea. The crests and ridges which compose it exhibit a few pigmies of the vegetable kingdom during the short interval of the melting of the snow. The principal of these are: *Androsace villosa, Aretia Vitaliana, Saxifraga oppositifolia, hyoides* and *muscosa, Antirrhinum alpinum, Iberis stylosa, Draba aizoides, Papaver alpinum, Potentilla apennina, Gnaphalium nivale, Gentiana nivalis.*

10. *Glacial region.*—This name is applied by M. Tenore to a few isolated points of the highest mountains of the Abruzzi, where the snow seldom disappears entirely. Its lower limit is defined by the appearance of *Cetraria*

z

islandica, which is accompanied by *Draba cuspidata, Festuca Halleri, Artemisia mutellina, Lepidium alpinum, Cerastium glaciale, Ranunculus brevifolius, Anthemis Barrelieri, Papaver aurantiacum, Gnaphalium dioicum.*

These ten regions may be readily recognized in proceeding from the shores of the Adriatic towards the summit of Mont Amaro, by Pescara, Chieti, Roccamorice and Majella, or towards the summit of Gran Sasso, by the route of Teramo, Montorio and Pietracamela.

The subdivision into so many zones as are here given appears unnecessary, and is not adapted to assist in comparisons with other regions. It would probably be sufficient to distinguish five regions, viz.: 1. The *maritime* region, not occurring at the foot of the inland mountains of Switzerland, &c. 2. The zone of evergreen trees, extending mostly to about 900–1200 feet, including the second and third of the preceding regions. 3. The zone of the chestnut and northern oaks, comprehending the fourth or first region of woods, from 900–1200 to 3000–3600 feet. 4. The zone of the beech, corresponding to M. Tenore's second region of woods : and 5. The alpine region, which seems unnecessarily subdivided, yet might be considered as subalpine, in relation to the Swiss vegetation, up to 6000 feet ; the remaining tracts, including the 'upper alpine' and 'glacial' regions above mentioned, forming a true alpine region analogous to that of the high Alps.

The limits and characters of these botanical zones will apply, with few modifications, to the rest of the northern and all the central portions of the kingdom ; but the differences are much more marked in the south. Thus, the mountains of the Basilicate and those of Calabria want most of the plants of Samnium and the Abruzzi, or these occur in places comparatively much more elevated. *Saxifraga Aizoon, petræa* and *calyciflora* occur at 7000 feet on Monte Pollino, while they are met with below 5000 feet in the Matese, Gran Sasso, and the Majella. It is also remarkable that none of the other Saxifrages of the north are found on the most southern mountains of the kingdom, as *S. biflora, cæsia,* and *muscosa.* The same may be said of almost all the characteristic plants of the three alpine regions, which do not occur even on the loftiest of the southern mountains, with the exception of *Draba aizoides, Thlaspi saxatile* and *Alyssum montanum,* which show themselves on the highest peaks of Dolce Dorme in Calabria.

Yet *Iberis Tenoreana* grows languishingly and dwarfed on the same summits, while it flourishes luxuriantly on the plateau of Faito on Monte Lactarius, below 3500 feet. *Alnus cordifolia,* on the other hand, maintaining a weak and uncertain existence on this plateau, spreads out a magnificent vegetation covering a large extent of the mountains of the Basilicate and of Calabria.

The proximity of the coasts of the kingdom of Naples to those of Greece and Africa causes the catalogue of its vegetable riches to be increased by many plants belonging to those countries. For example, *Alyssum creticum, Cachrys Libanotis* and *triquetra,* and other plants of the Greek flora, grow upon Monte Gargano, on the shores of the Adriatic and of the Ionian Sea; while *Spartium villosum, Rottböllia fasciculata, Sinapis radicosa,* and several other species belonging to the Atlantic flora, extend as far as the environs of the capital. Even examples of oriental plants are not wanting on the western coast of the kingdom, as *Anthemis Chia,* found by Tournefort only in Scio, but now growing near Reggio, on the west coast of Calabria, where moreover the myrtle, the laurel and oleander, with their evergreen foliage, agree wonderfully well with the poetic character of this classic soil. On the other hand, none of these trees grow spontaneously in the northern region, and they are only cultivated with difficulty.

This diversity of vegetation has caused totally different systems of cultivation to prevail in the two regions. The olive is not cultivated in elevated parts of the outer Abruzzi; neither the orange nor the citron resist the cold of the climate of the inner Abruzzi and Samnium; the mulberry does not prosper there, and the vine yields a poor wine; while dense and odoriferous groves of citrons and oranges embellish the environs of Reggio, and all Calabria produces silk and exquisite wines. On the other hand, saffron is cultivated with the greatest success in the outer Abruzzi.

The sugar-cane was cultivated up to the 17th century on the coasts of the Ionian Sea and in Calabria, and the produce was sufficiently abundant to allow of considerable exportation. But all the recent attempts to introduce this culture in the environs of Naples have been unsuccessful; and the specimens cultivated for curiosity in the Royal Garden have to be kept in the orangery all the winter. The same precaution is necessary in Naples with *Musa paradisiaca, Acacia Lebbek, Gossypium arboreum, Anona tripetala, Brugmansia arborea, Ficus elastica,* and other similar plants; but it is most probable that they would succeed in the open air at Reggio, as well as they have done at Palermo, where the temperature of winter does not differ from that of the coast of Calabria. But at the same time, the climate of Naples allows of the growth in open air of the *Camellia, Metro-*

sideros, Melaleuca, Eucalyptus, Banksia, Laurus Cam-phora, Acacia falcata and *longifolia*, with many others, natives of the Cape, Japan, and New Holland, culti-vated in stoves in almost all other parts of Europe, and which cannot sustain the heat of the height of summer in a country hotter than Naples, where, on the other hand, the Rhododendrons, Kalmias, Azaleas, and similar plants of the north of Europe cannot be preserved.

A very interesting fact was announced by M. Tenore in his 'Flora of Naples'; namely, the discovery of *Pte-ris longifolia* and *Cyperus polystachyus* close to the *fu-marolles*, or smoking craters, of Frasso and Cacciotti, in the island of Ischia, vegetating in earth, the temperature of which never descended below 77°, on account of the heat exhaled from them; and the intensity of this heat was such, that in digging to the roots of the plants the hand could not be held in the hole without being burnt. *Pteris longifolia* is indigenous in Jamaica and New Spain; *Cyperus polystachyus* grows in the E. Indies, Arabia, and N. Africa. M. Tenore attributed the presence of these plants to their having persisted from a former geo-logical epoch of the globe, from the time when palms and other tropical trees grew in Europe. It seems more likely that they had been recently introduced.

A striking example of the crowding together of plants of different climates is obtained in passing from this island of Ischia to Castellamare; where, near the chapel of St. Ange, on the summit of Monte Lactarius, may be gathered *Cerastium latifolium*, a species regarded by botanists as indigenous on the highest Alps, while on the steep peaks of the same mountain grow *Rhamnus pusil-lus, Pedicularis foliosa*, and *Saxifraga Aizoon*.

Thus, under the same parallel, and in a line of at most

thirty miles, we may gather the plants of the most distant
countries of the earth! A combination probably unique,
and of which, at all events, no second instance is at present known.

In Sicily are found all the circumstances favourable to
the combination of the vegetation of the south with that
of the north. There the sugar-cane, banana, anona, date,
&c. are cultivated with more or less success. The hedges
are composed of the agave, or American aloe; and beside the plane, the poplar and the willow, grow the
cactus, orange, citron, olive, myrtle, laurel, St. John's
bread, pomegranate, &c., while the arbutus and tamarisk
abound on the shores. The dates of Girgenti, situated
on the south coast, are very good, but at Palermo the
tree does not develope its fruits freely.

Of all the phænomena of vegetation, however, presented by Sicily, those of Etna are the most interesting
and varied; and since this is the most southern of the
great summits of Europe, it demands especial attention
in respect to the relations of its zones of altitude to those
of the mountains in the centre and north of the continent.

The vegetation of Sicily exhibits only four well-marked
regions :—1. The region of maritime or shore plants;
2. The cultivated region, limited above by the cessation
of the culture of the vine, from 0 to 3300 feet; 3. The
wooded region from 3300 to 6200 feet, beyond which no
trees are found; and 4. The subalpine region, which
terminates above in a bare and lava-covered surface, exhibiting but few cryptogamic plants, and scarcely forming an analogue to the alpine regions of the Alps, Pyrenees, &c.

1. The *maritime region* is hardly at all developed at

the foot of Mount Etna, since this mountain falls by rude lava-cliffs into the sea; the sandy dunes and plains in which most maritime plants delight occur first to the south of Catania, and do not belong to Etna.

2. The *cultivated region* commences immediately above the sea, and extending up for 3300 feet, exhibits a fertility and beauty which excite the enthusiastic admiration of every traveller. Here the gardens exhibit a crowd of tropical plants; the banana ripens its fruit; *Erythrina corallodendron, Hibiscus mutabilis, Cassia biflora, Datura arborea*, and *Cæsalpinia Sapan* display their large and brilliant flowers; the date gives the landscape an African aspect; while the *Cactus Opuntia* and *maximus*, often 12 feet high, and the *Agave americana*, which frequently sends up a vast pyramid of blossom in its third or fourth year, give a strange and completely southern character to the scenery.

Little grain is cultivated in the lower region of Etna, the soil being mostly rocky and unfit for it. Barley is the common food of the cattle, both as corn and as green fodder; oats are almost unknown in Sicily, and the horses feed now upon barley as in the days of the siege of Troy; wheat is cultivated only up to 1600 feet, but this does not seem to be a limit given by nature; maize is little grown in Sicily, and scarcely at all on Etna. Besides the ordinary European esculent vegetables, among which gourds, peas, beans, and lupins are conspicuous—(the two last especially as the common food of the poor),—many fruits are abundantly grown, particularly figs, almonds, peaches, apricots, pomegranates, and pistachio nuts. The walnut is very rarely seen; the hazels, *Corylus Avellana* and *Colurna*, are grown to a considerable extent, especially for exportation to England. Near water flourish the southern fruits of the orange tribe—the orange, citron, lime, lemon, &c., in the greatest abundance and countless varieties. They succeed very well up to 1850 feet, excepting certain cultivated varieties; on Nicolisi (2184 feet) they are affected by frost in severe winters, and their limit may be taken at about 1900 feet. The date does not rise above 1680 feet, and its fruit has never much flavour; still in 1832 there was a tree 10 feet high in the botanic garden of Palermo, only 14 years old, raised from Sicilian seed. The fig bears good fruit up to 2200 feet; the stone pine, as in the kingdom of Naples and Sicily generally, occurs only solitary and cultivated. The sugar-cane is not cultivated in the gardens on Etna, nor the tanner's sumach (*Rhus Coriaria*), which is so largely and increasingly grown in other parts of Sicily. Much cotton, on the other hand, is raised on the banks of the Simeto, and is found up to 1000–1200 feet. The great Italian reed (*Arundo Donax*), the tree-like shaft and broad leaves of which remind us of the tropical bamboo, is much grown for stakes for the vines, and many other purposes; it ascends to 2500 feet. The black mulberry, on which the silkworms are almost exclusively fed, goes as high; the white mulberry is much more rare.

The cultivation of the olive is pretty general at the foot of Etna, up to 2200 feet; but the greatest proportion of this region is occupied by vines, for which the soil is so pre-eminently fitted. They are, almost all over Sicily, trained to low stakes of the giant reed, and never on trees as in Lombardy, Naples, &c. The highest points to which this culture extends are Portella, near Zaffarana, 2973 feet, and the slope of Monte Zoccolaro, 3300 feet.

On the roughest grounds, covered with fragments of lava, flourish the opuntias even up to 2200 feet, often forming large thickets, growing where nothing else will, and furnishing abundance of succulent fruit highly valuable in such a climate ; the cattle and goats also feed upon the juicy stems. Three species are cultivated in Sicily : *Cactus Tuna, C. Opuntia*, which yields the edible fruit, and *C. maximus* ; there are many varieties of the opuntia or prickly pear, with red and green fruits, the latter most prized for their delicate aromatic flavour.

Under the shade of the cactus flourish numerous wild plants, such as species of *Lupinus, Calendula, Asphodelus, Asparagus, Silene, Brassica, Sinapis, Reseda fruticulosa, Acanthus mollis, Arum Arisarum*, &c., which, without this protection, would be burnt up by the torrid rays of the summer sun. These and the *Cacti* gradually form a vegetable soil by their decay, which subsequently becomes capable of culture.

The other plants most abundant on the lava streams are :—*Andropogon hirtus* and *distachyos, Lagurus ovatus, Rumex scutatus, Valeriana rubra, Plumbago europæa, Thymus Nepeta, Satureia græca, Ranunculus bullatus, Capparis rupestris, Alyssum maritimum, Isatis tinctoria, Scrophularia bicolor*, several species of *Linaria, Heliotropium Bocconi, Mandragora autumnalis, Prenanthes viminea, Apargia fasciculata, Senecio chrysanthemifolius, Daphne Gnidium, Spartium infestum* and *junceum, Physalis somnifera, Solanum sodomæum, Ricinus africanus, Smilax aspera, Euphorbia Characias* and *dendroides*, the tree-like Euphorbia, one of the handsomest shrubs of Sicily when in blossom in spring. The mandrake often clothes great tracts with a blue carpet of its countless flowers in autumn ; a few species of *Sedum* occur upon the rocks, but not a single *Sempervivum* (unless we include *Sedum rostratum* in this genus). The principal ferns are those of dry exposed stations.

3. The *wooded region.*—The wooded region is pretty distinctly defined below at about 3300 feet, but is said to have been more extensive formerly ; destruction having been caused both by eruptions of the volcano and by the reclaiming of the land to cultivation. The chestnut does not appear to be wild on Etna ; it has been observed as high as 3900 feet, while in the Pyrenees it only attains 2800 feet, and in the Alps 2500 feet. The old chestnuts of Etna are widely famed ; the *Castagno di centi cavalli*, as it is called, is 180 feet in circumference near the root ; the *Castagno di Sta. Agata*, 70 feet ; the *Castagno della nave*, 64 feet. Their trunks are not very high, but branch at a short distance from the ground ; Dr. Philippi states it as his opinion that the *C. di centi cavalli* is not one single trunk, but several from the same root, for it has at present five quite distinct trunks, and it is a common custom to cut down the chestnut when it is about 1 foot in diameter, and then it sends up a number of stems from the crown of the root.

The woods of Etna consist chiefly of *Quercus pubescens,* Willd., which is also the principal constituent of the woods of the Apennines, at least in Southern Italy ; it extends from 3200 to 5500 feet in one place ; on the east side only to 5100 feet. *Quercus Cerris* is common in the Val del Leone, but does not extend higher than 4600 feet. The evergreen oak (*Q. Ilex*) ascends from the low hills of the coast, where it is the prevailing tree, to 3800 feet in the Val del Bue. The beech is not found below 3000 feet, and rises as high as 6000 feet. The birch (*B. alba*), which does not occur throughout the whole range of the Apennines in the kingdom of Naples, excepting at the southern extremity, where it does not reach 5600 feet, forms little woods as high as 6100 feet on Etna, never occurring below 4760 feet. A pine, which has been called by some authors *sylvestris*, by others *Laricio*, is frequent on the east side of Etna, forming stately trees 120–130 feet high; the lowest point at which it has been observed is 4000 feet, the highest 6200 feet. The aspen occurs sparingly at 5500 feet, the holly at 4760 feet, neither of which often exceed 4600 feet on the Alps.

The handsome broom, *Genista ætnensis*, is peculiar to this region, and where cultivated becomes a tree, resembling somewhat when out of flower, with its long, slender, leafless and pendent branches, the casuarinas of New Holland; wild, at 3980 to 6000 feet, it is always shrubby, and then does not assume that foreign aspect. *Daphne Laureola* is common; *Erica arborea* infrequent. Towards the upper part of this region, *Juniperus hemisphærica, Astragalus siculus, Berberis vulgaris* var. *macroacantha* occur, but they belong rather to the next region. The common European orchard fruits succeed best in the lower part of the wooded region, the climate being too hot lower down. No other grain is cultivated here but rye, which was introduced in the beginning of the last century from Germany.

This region displays great poverty in species. The soil beneath the trees is densely covered with the common brake fern, which in many places drives out every other plant. It is commonly accompanied by *Crocus odorus* and *Cyclamen neapolitanum*; other characteristic plants are: *Sternbergia lutea, Asphodelus luteus, Potentilla calabrica, Gypsophila rigida, Centaurea cinerea,*

Achillæa ligustica, Tolpis quadriaristata, Apargia his-pida and *autumnalis, Thymus Acinos, Satureia græca, Paronychia hispanica, Herniaria macrocarpa,* &c. *Croton tinctorium* has been observed as high as 5090 feet. This region also contains all those found in the next.

4. The *subalpine region,* 6200 to 8950 feet.—*Juniperus hemisphærica* and *Berberis vulgaris* rise out of the preceding region and ascend to 7100 feet in this. The vegetation of this region acquires a peculiar aspect from the *Astragalus siculus,* which is the prevailing plant here, and perhaps to a certain degree represents the rhododendrons of the Alps and the *Spartium nubigenum* of the Canary Islands. It forms dense hemispherical tufts 4 or 5 feet in diameter and 2–2½ feet high, covered with prickles; it rises as high as 7500 feet, as does also *Tanacetum vulgare,* which is tolerably frequent at 3000 feet in the wooded region. Higher than this all shrubby plants vanish. The few scattered plants occurring in the black sand, which is spread far and wide over the summit of Etna, are chiefly : *Saponaria depressa, Cerastium tomentosum, Cardamine thalictroides, Viola gracilis* or *ætnensis, Galium ætnicum,* .*Sesleria nitida, Scleranthus marginatus, Seriola uniflora, Anthemis punctata,* and *Rumex scutatus;* the last is common on all the lava streams, and descends to the level of the sea, the only change at great altitudes being that its leaves become more glaucous and hairy. The last plant is *Senecio chrysanthemifolius,* with its varieties *incisus* and *carnosus,* at 8850 feet, and with it we lose the last trace of vegetation, although no snow lies upon the summit in the summer, and from hence to the peak is found a barren waste of lava and ashes, without a trace of life, except the footmarks or perhaps the bones of mules, which

show that a desire of knowledge or of novelty has often driven human beings up to the barren peak. Dr. Philippi states that he found no trace of the Lichen region from 9000 to 9200 feet, described by Presl; and the higher parts of Etna generally are very poor in Cryptogamous plants.

The most striking feature of the vegetation of Etna is the paucity of species and of individual plants in the wooded, and more particularly in the subalpine region, contrasting so strongly with the abundance and variety of the plants of the Alps and Pyrenees. The principal cause of this is to be found in the extreme dryness of the soil and climate; the summit of the mountain is rarely enveloped in clouds, rain falls but seldom, and there is no accumulated mass of eternal snow, which by its slow melting under the summer's sun should constantly irrigate the declivities. No springs are found upon the slopes; the little water that is precipitated upon Etna only comes to light low down, where the lava rests upon the original argillaceous subsoil.

Another important cause is the peculiar nature of the soil. In the Alps the primitive and stratified rocks have been exposed for ages to all the influences of weather, and no interruption has checked the gradual production of fruitful soil and the multiplication of vegetable life upon it. But the case has been very different on the higher parts of Etna. A fresh lava stream, a field of new ashes, is not readily clothed even with a sparing coat of vegetation; and where this in time comes to pass, a new eruption in a moment destroys the slow product of centuries. It is calculated that there have been fifty-four eruptions of Etna during our historical period, that is, one about every thirty-three or thirty-four years, so

that the surface of the upper part of Etna may be regarded as only of historical antiquity. It has probably,
therefore, derived its plants from the neighbouring regions; but these are plains or mountains of not more
than 3000 feet, therefore we should only expect to find
on Etna plants proper to such stations. These can only
rise to a certain height, since they require a certain
amount of heat and other conditions which are no longer
met with at great altitudes. The plants proper to a cold
climate, prevailing on the higher parts of Etna, could
only be derived from the Alps, the higher mountains of
the Abruzzi, or of Calabria which do not exceed 7200
feet. We should consequently expect but few such
plants, on account of the difficulties of communication.

These theoretical conclusions are fully confirmed by
experience. The species which have been detected above
the tree-limit on Etna amount to only 47, and when
these are compared with the inhabitants of the Alps or
those of the higher peaks of the Abruzzi, which closely
resemble the latter, it is found that the vegetation of
the summit of Etna is quite distinct in its character.
Of all the 47 species, there is but one which is not
found both over the rest of Sicily and on the neighbouring part of the continent; only two not in Sicily, but in
Calabria. All the rest are common to all Sicily, and
most of them belong to the lower, warmer region; and
there appears to be only one plant, *Genista ætnensis*,
peculiar to Etna.

A striking phænomenon is seen in contrasting the limits
of different plants on Etna and the Alps. The limits of
the Cerealia, the olive, chestnut and beech, have a determinate and equal relation; they are all 1300 or 1400
feet higher on Etna than on the south side of the Alps.

But the limit of the growth of trees is not at all higher. This apparent anomaly, however, is not dependent on climatal conditions, but on the continual changes of the surface of the upper parts of Etna by eruptions, preventing the accumulation of vegetable mould capable of supporting trees. The limit of the culture of the vine again is relatively low upon Etna, but it is evident that this depends on mere economical causes; it is not worth while to labour for the production of an inferior article at great heights, when so rich and fertile a country lies around.

Like those of the subalpine region, the plants of the wooded region differ much from those of the analogous region of the Alps. There are no species of currant, *Vaccinium, Pyrola*, aconite, saxifrage, gentian, &c.; even the wild strawberry is a rarity. Neither does the vegetation resemble that of the neighbouring mainland of Italy, nor even that of other higher mountains of Sicily; for on the Madonians, for instance, many subalpine species occur which are altogether wanting on Etna. There is a considerable resemblance,—bearing in mind always that the oaks, beeches, and hornbeams of Etna are replaced by laurels and *Myrica Faya*,—to the wooded zone of Teneriffe. The highest tree is a pine in both cases, and the brake fern is extremely abundant on both; of 90 plants found in this region in the Canaries, one-fifth occur also on Etna.

There are some peculiarities about the distribution of the trees; the distinctions seen in other parts of Europe seem to be lost here. On Mount Etna, the beech, the birch and the Scotch fir occupy the same zone. In the Pyrenees the beech ceases before the Scotch fir begins, and in the Alps the birch fails before the Scotch fir.

But in Scandinavia the birch extends far above the Scotch fir, and in fact ascends higher on the mountains than any other tree, as it also goes farther north in Russia. There seem, therefore, to be peculiarities in the constitution of these trees, which modify the general law of the relation of altitude to latitude, and appear to show that an *alpine* flora is not necessarily *arctic* in its character.

The plants of the cultivated region are those of the neighbouring kingdom of Naples. A few are added for which the climate of the north of Naples is too cold. Among these are the *Solanum sodomæum*, or apple of the Dead Sea, the oleander, and the *Ricinus africanus*, which is here and there arborescent. The dwarf palm, so common on the south and west sides of Sicily, does not occur in the Etna district. There is no resemblance whatever to the vegetation of the south foot of the Alps ; but a certain amount to that of the *Mediterranean region* of the Canaries, for of 140 species stated to occur there, 65, or nearly half, occur in Sicily.

Sect. 10.

The Greek Peninsula.

North and South Europe are separated by a vast and lofty mass of mountains, running from E. to W. in a curved line, only interrupted at one point. This great barrier is composed of the Balkan, the Dinaric Alps, the true Alps, and the Pyrenees.

The characters of the Balkan have not been so thoroughly investigated as those of the other ranges.

Their height is considerable, and they extend from the Black Sea nearly to the Adriatic, between the valley of the Danube and the Greek peninsula, about eighty geographical miles from E. to W. The snow-covered Orbelos is regarded as the highest point, being estimated at 9500 feet.

The Balkan is connected with the German Alps by a chain seventy geographical miles long, running from S.E. to N.W., parallel to the Adriatic, which is known by the name of the Dinaric Alps. Mont Dinario appears to be the highest summit in this range, rising 7396 feet above the sea; the Klek attains 6906 feet. These mountains are for the most part in the neighbourhood of the coast of the Adriatic, and fall very abruptly towards it; the islands lying off also, such as Cherso, Grossa, &c., are mountainous, and reach a height of 3000 feet and over. The mountains are calcareous and the soil generally very dry. The vegetation of the S.W. declivity agrees very closely with that of Italy and Greece; on the N.W. slope they share the character of the Hungarian and East-European plains.

The winter of Dalmatia is very mild, and vegetation awakens in February, the almond flowering even in January; the active season lasts till May, when a dry, hot summer commences, continuing till the end of September, during which the annuals disappear and the perennial vegetation is parched. The rainy season of September and October follows, in which many plants flower for the second time, as in Spain and Italy.

Dalmatia is said to comprehend three well-marked zones of vegetation, viz.

1. The coast region, from the Adriatic up to a level of 1400 feet, characterized by the presence of the olive,

arbutus, laurel, oleander, stone and Aleppo pines, Cisti, lentisks ; in short, the usual warm Mediterranean flora.

2. This is surrounded by a hilly region, extending from 1400 to 3000 feet, on which the general aspect of vegetation assumes more of a Central-European character, but with the species of that flora are intermingled some southern and eastern forms. This is the region of the oak (*Q. Cerris*), beech, sycamores, &c. &c. Among its other characteristics are *Cytisus Weldeni, Gentiana lutea, Valeriana montana* and *tripteris, Centaurea montana* and *tuberosa, Helleborus multifidus, Primula suaveolens,* &c. Some *alpine* forms occur at a low level in this region, as is the case on the coast of Scotland.

3. The subalpine region rises to 6000 feet, and presents certain shrubs, the dwarf juniper, *Rosa alpina* and *Lonicera alpigena,* with the herbaceous *Dryas octopetala, Arabis alpina, Androsace villosa, Pæonia Russi,* various Silenes, Campanulas, Arenarias, &c.

The proper peninsula of Greece itself, differs considerably in form from the other peninsulas of Southern Europe ; it diminishes in breadth towards the south, is indented by the sea in various places, and terminates in a smaller peninsula, the Morea, only connected with the mainland by an isthmus. The northern boundary is formed by the Balkan and Dinaric mountains, as the Pyrenees and Alps respectively bound Spain and Italy. Numerous islands surround the other three sides, among which Candia, Negropont and the Ionian Islands are the largest. Greece lies between the 41° and 36° of latitude ; Candia stretches to the 35°.

The greater portion is mountainous ; in the north, branches of the Balkan and Dinaric Alps run in ; the southern portion (Greece proper) appears to consist in great part of several detached masses of mountains, which in part form table-lands. The most important in extent and height are:—

	Feet.
Pindus	6906
Taygetus, in the Morea	7914
Ida, in Candia	7650

The other islands are also mountainous.

Greece has no rivers of considerable size; the most important are the Maritza, Vardar, and Drio, which spring from the Balkan and flow southwards. Many lakes occur, but none of great size. Limestone appears to be the prevalent rock, especially in the west, as in Italy and the Dinaric Alps; some of the islands exhibit traces of volcanic products, and from the mountain of Calamo in Milo, and from the little island of Kameni near Santo, sulphurous vapours are evolved, but no lava-streams occur.

This peninsula is colder than the Italian and Spanish; this is especially the case on the east side. Constantinople, although lying in the latitude of Naples (see Map), has no oranges, not even the olive; and frosty nights, with snow, are not uncommon. The climate is indeed milder in the south of the peninsula and in the islands, but still colder than in the west in the same latitudes; for example:—

	Lat.	Ann.	Wint.	Sum.
Canea (in Candia)	$35\frac{1}{2}°$	$65°$	$54\frac{1}{2}°$	$78°$

So that the mean temperature is 1° only higher than Palermo, $2°\frac{1}{2}$ more to the north. The temperature is also low upon the table-lands, as in the Morea and near Yanina. The quantity of rain seems to be less than in

Italy; the summer rains are slight here also. The mountains of Greece do not reach the snow-line, but snow is almost constant on the higher mountains of the north of Greece, and the most southern of its plains are subject to frosts. In the Morea, near Tripolitza, the thermometer sometimes falls 16° or 18° below the freezing point; but snow is rare, and of short duration, in the peninsula, except on elevated situations, where it lies throughout the winter.

There can be no doubt, however, that various local causes have influence on the climate of Tripolitza, since the Morea produces the orange, citron, and even the opuntia, which is scarcely less susceptible of cold than the date: this spiny plant, so abundant in most parts of the Mediterranean region, is used for hedges in the fields of Messenia. Apparently there are no dates in the Morea, and a few individuals in the neighbourhood of Athens seem to be the only examples that exist on the mainland of Greece.

On the eastern coast the orange and citron penetrate by Bœotia, Phocis and Thessaly to Mount Olympus, which separates Thessaly from Macedonia. Beyond this point the country is full of mountains, exposed to violent winds, and it is covered therefore with forests resembling those of Central Europe. Neither the orange nor the citron is found in the famous Vale of Tempe on the south side of Olympus; and although they grow on the island of Lemnos, in the same latitude, their fruits do not ripen. The olive flourishes on the coast of Macedonia up to 41° north latitude.

Judging from the vegetation, the west coast is warmer than the east. Near Epirus, between 39° and 40° N. L., in exactly the same latitude as the Vale of Tempe, Corfu, celebrated for its fertility, produces both the opuntia and the date.

The plants belonging to the zone of transition between the tropics and the temperate zones extend from Epirus into the Illyrian provinces. The olive and myrtle, the orange and citron, ornament the romantic rocks of the mouths of the Cattaro and the shores of the Gulf of Guarnero. The orange and citron do not go beyond this point; the myrtle and laurel, with the southern oaks, *Quercus coccifera, Ilex* and *Ægilops,* the oriental hornbeam, the manna-ash, the stone pine, pistachio, caper, &c., follow the coast region as far as the upper end of the Adriatic. But inland this vegetation is suddenly arrested at a short distance from the coast by the mountains, which present the plants of temperate Europe.

Greece possesses few large species characteristic of the Mediterranean climate which it does not share with the other parts of the region. The most common trees and shrubs from Cape Matapan to Olympus, on the east, and to the southern frontier of Dalmatia, on the west, are: the olive, *Jasminum fruticans, Phillyrea media* and *angustifolia, Styrax officinale, Arbutus unedo* and *Andrachne,* the myrtle, the pomegranate, the bay-laurel, *Ceratonia siliqua, Cercis Siliquastrum, Pistachio terebinthus* and *Lentiscus, Ziziphus vulgaris* and *spina Christi, Paliurus australis, Rhamnus Alaternus,* the caper, *Acer monspessulanum,* the laurel, *Osyris alba,* the fig, *Celtis australis,* several poplars, including the Lombardy and the aspen, the cypress, the stone pine, *Juniperus phœnicea* and *macrocarpa,* with the savine, several *Cisti,* &c. &c. On the banks of running waters or in damp places grow the Oriental plane, many willows, the alder, *Vitex agnus-castus,* and oleander; on the sea-shore, the stone and pinaster pine, *Quercus Ægilops*. On the mountains, *Abies taxifolia*, the hornbeam, *Salix retusa* (these three in the highest regions), the Scotch fir, the yew, northern oak (*Q. Robur*), *Ostrya vulgaris*, the beech, chestnut (in the middle altitudes of the mountains), the hazel and *C. Colurna*, the aspen, the manna-ash, the limes, the horse-chestnut, the mountain-ash, with the crabs, wild pears, service-tree, &c., the evergreen oak (*Q. Ilex*), with *Q. ballota* and *coccifera*, the three last preferring the lowest valleys, or even the plains.

The hills of Istria are clothed with the vitex, pistachio; jasmine, myrtle, fig, olive, pomegranate, &c.; while the cypress, the evergreen oaks, manna-ash, stone pine, sumach, caper, laurel, &c., with many of the annual or perennial herbs of the Mediterranean flora, extend to the vicinity of Fiume and Trieste.

The general character of the vegetation thus resembles that of Italy, and the cereals are the same. Cotton is cultivated extensively, and goes as far north as the Dardanelles on the east side ; the orange does not succeed so far up. The cultivation of tobacco is also considerable.

The following statistics relative to the flora of the Morea will illustrate some of the relations to the other parts of southern Europe.

Of 1550 plants enumerated by Bory St. Vincent, 1356, belonging to 97 families, are flowering plants. Suppressing the oranges, Meliaceæ, *Cacti*, Sesameæ, Eleagneæ and palms, which are not indigenous, the flowering families of the Morea and the Cyclades are reduced to 91. Of these the Cucurbitaceæ only include 2 indigenous species ; 13 other families only contain 2 species each, and 21 only 1 species. The richest families are, taken in the order of their numerical importance, the Leguminosæ $\frac{1}{9}$, Compositæ $\frac{1}{10}$, Grasses not quite $\frac{1}{15}$, Labiatæ $\frac{1}{20}$, Cruciferæ rather less than $\frac{1}{20}$, Caryophyllaceæ $\frac{1}{24}$, Umbelliferæ $\frac{1}{27}$, Ranunculaceæ rather more than $\frac{1}{35}$, Cyperaceæ and Liliaceæ $\frac{1}{39}$, Orchidaceæ $\frac{1}{40}$, Boraginaceæ $\frac{1}{42}$, Scrophulariaceæ $\frac{1}{43}$, and Rosaceæ rather more than $\frac{1}{44}$.

The Monocotyledons amount to about 260, the Dicotyledons to about 1085. Among the largest genera are those of the Leguminosæ, as for example: *Trifolium* 25 species, *Vicia* 16, *Medicago* 13, *Lathyrus* and *Lotus* each 12. There are 20 *Euphorbiæ*, 19 *Silenes*, 18 *Gerania*, and the same number of *Ranunculi*; *Carex, Bromus, Carduus, Centaurea, Orchis, Cistus, Veronica* and *Rumex* have each about 16 species.

The flora of the Morea is closely related to that of Italy and the south of Spain, the European forms which it has in common with them being intermingled with African forms, belonging to the Libyan and Barbary floras. On the whole it presents no very striking features, and the general aspect is very similar to that of Provence. As usual in these southern regions, there is rarely any verdant turf to be met with, such formations being confined to the mountains or more shaded valleys. The Morea seems to possess much less wood than formerly, and even that is disappearing.

The islands on the west of Greece form a transition to the Italian coast, and Zante may be selected as an example of the character of the Ionian Islands. Situated in the 37° 45′ N. L., Zante is elongated in a direction parallel to the coast of the Morea. It is traversed lengthways by a range of calcareous mountains, of a cavernous limestone, dividing it into two perfectly distinct portions; that on the west dry, hilly, and little cultivated; that on the east, facing Greece, distinguished chiefly by a low plain nearly level with the sea, and enriched by careful cultivation.

From Cape Skinari, on the N.E., this chain rises gradually, runs south, attains 2200 feet at Mont Varchiona, the highest point of the island, and drops suddenly down to the sea, where it forms the cape and bay of Chieri. The western slope forms a kind of plateau inclined towards the coast of Italy, terminating at the seaside in low hills or high cliffs. Open to all the force of the winds, stony, irregular, and devoid of water, this plateau presents a woody, hard, and apparently scanty vegetation. *Globularia Alyssum* attaches itself firmly to the rocks, and covers large spaces; *Phillyrea media* is only an under-shrub, scarcely a foot high on the most exposed hills; but in the damper valley of Spiliotissa, *Erica arborea* grows to a height of 4 or 6 feet, side by side with *Erica verticillata*, which also occurs abundantly on other parts of this plateau.

Thirty villages occupy the less rocky valleys of this slope, forming so many oases, where industry has made even this ungrateful soil productive; they have around them corn-fields, vegetable gardens, and even vines. *Narcissus Tazetta* is very abundant wherever there is shade and moisture; *Fritillaria plantaginea* covers entire fields. The cultivated olive has displaced the wild trees, and the latter are restricted to certain localities, where they are said to be preserved from a religious feeling. *Pinus Pinea* and *Pinus halepensis* surround and protect some of the buildings, and they would form a forest around Cape Skinari in the course of some years, were it not that they are annually sacrificed for the purposes of the moment.

The eastern declivity descends by steep slopes into the low plain, and many villages exist upon them ; communication is carried on by means of roads, or rather paths, impassable by carriages ; here the rocks are bare of turf, everywhere composed of perforated limestone, worn by the action of water, and rolled down in fragments among the thickets. Otherwise the vegetation does not present any remarkable difference from that of the hills on the other side. *Salvia triloba* grows from the base to the summit, with *Poterium spinosum, Orchis rubra,* and *Putoria calabrica. Iris unguicularis* also rises to the summit of the ridge, with *Inula viscosa,* which covers the whole country in summer. *Narcissus serotinus, Crocus Boryi* and *Lithospermum orientale* seem to be peculiar to this region.

On the shore facing the Morea there exists a tolerably uniform system of stratification. Starting from the north runs a chain of low hills, sending out little tufaceous promontories into the sea, which seem most frequently to rest on a bed of plastic clay ; this bed extends further into the island, and appears to be the cause of the retention of the moisture which nourishes the roots of the vine during the four or five absolutely dry months of summer. More to the south the shady hills of Acrotiri descend abruptly to the shore, and display almost perpendicular cliffs of a blue clay, surmounted by tufaceous layers carpeted with *Samolus Valerandi* and ferns ; near the town this tufa is clothed with *Silene gigantea* and beautiful groups of *Chamæpuce mutica,* which forms a shrub with a stem 3 or 4 inches thick. These hills join the fortress, which itself is but a mass of clay and gravel, about 500 feet high, characterized by great abundance of *Thapsia garganica, Evax pygmæa, Eryngium creticum* and *virens, Carduncellus cæruleus,* &c. Here the chain terminates, and the principal stream of the island discharges itself into the sea. Still further to the south rises the isolated calcareous hill of Scopo, the north-eastern slope of which, watered by several brooks, presents cultivated spots and a more abundant vegetation. The S.W. side is dry and almost perpendicular, while on the S.E. side the mountain rises by a gentler inclination to the low tufaceous hills forming the southernmost point of the island, Cape Geraka. The tufa also predominates in the structure of the isle of Peluso, where *Ononis Cherleri, Coniera implexa, Anacamptis pyramidalis, Elichrysum rupestre,* &c. are found.

Mount Scopo appears to bear upon itself a large portion of the vegetation of the island. The plants of the sea-sands, in cultivated soils and inundated places, grow at its foot ; most of those of the other mountains occur on its sides : some species seem to be peculiar to it, as *Silene italica,* which grows at the summit with *Carduus tenuiflorus, Bunium junceum, Lagœcia cuminoides, Nigella damascena,* an oak, supposed to be *Q. Ilex,* forming a little wood near the summit, *Helianthemum guttatum, Sideritis purpurea, Hypericum empetrifolium,* and more rarely *Scorzonera crocifolia* and *Cytisus hypocystis,* two species only found there.

The climate is much the same as that of the islands of the Archipelago in the same latitude. The variations are regular : towards the middle of April, when the winter and spring rains are quite over, the sky becomes bright and remains so for four months ; the earth has been sufficiently watered to last through the summer, and unfolds all its riches. In the middle of May the fodder plants are mown. In the middle of June commences the harvest of cereals, and from this time the temperature rises towards its maximum, which is generally not below 84°, but varies between 86° and 100° in the hottest days of July and August. This heat is accompanied by the *sirocco*. In October the rainy season usually commences, that is, after the gathering of the Corinth grapes; but it is not unfrequent for temporary rains to fall in August and September, which are very injurious, since the fruit, being ripe, is caused to rot by a rain lasting a few hours. October and November are stormy months; rain is almost constant in December and January; snow is rare; it is said that sometimes twenty-five years pass without it.

The cultivated land forms about four-tenths of the total surface, principally between the mountain-chain and the hills above spoken of, and upon those hills, where the olives and oranges present perpetual verdure. The low plain of the east, with its hills, has made the reputation of Zante by the fertility and picturesqueness. The soil, usually a mixture of sand and clays, affords to the vines, and above all to the Corinth grape, a favourable soil, which renders their culture profitable. The Corinth grape appears to be a native of Asia Minor.

The plain, which in some places resembles an immense vineyard, exhibits also a number of arborescent plants. The cypress grows there, though not wild; the quince, either cultivated or run wild in the hedges; the elm (*U. campestris*), rarely; while four species of *Citrus* are cultivated in the inclosures—the citron, the lime, the orange, and the lemon. The last two are extensively cultivated.

There are also tolerably large plantations of figs of different varieties. The common fruit-trees of Europe, imported from Germany or Italy at various periods, are also found in cultivation.

The plantations of vines are very often accompanied by the *Agave americana,* used as a fence, as also in the fields and sometimes in the gardens. It is found all over the cultivated parts of the islands, often with *Cactus Opuntia,* which however is much less common.

The true arborescent vegetation of Zante consists of its olives, which are cultivated in extensive woods in the lower portion, on the eastern hills, at the foot of the mountains, and in some localities on the plateau.

Corn-fields bear a small proportion to the other cultivated lands. Some places are devoted annually to the growth of the cotton (*Gossypium herbaceum*); other spots are sown with *Sesamum orientale,* the seed of which is used as an article of food throughout Greece. *Cicer arietinum* is cultivated in the same way. A little maize is found around the farms; wheat is grown in the plain, on the plateau, and even among the olives, but only in sufficient quantity to serve the inhabitants three or four months out of the year; barley and oats are still more neglected, since the natives prefer employing the arable soil in the growth of the vine, and depending on Greece or Odessa for their grain.

There are no natural pastures. The ordinary fodder consists of trefoil, lucerne, &c., which are grown in the olive groves where the soil is good and the trees far enough apart. The name of a natural meadow might perhaps be given to one marshy plain, covered with water during five months of the year; its borders are solid and firm, and formed of close turf, which serves to pasture the cattle; but the greater part of the numerous sheep and goats existing on the island browse on the grasses among the thickets of the mountains, and along the hedges and roads.

With regard to the botanical relations of Zante, out of 626 flowering plants found by M. Reuter, 455 are met with in the kingdom of Naples; 429 in Sicily; 370 are common to Zante and those two countries; 527 are indicated for the Morea and the islands of the Archipelago; and among these 527, there are only 29 which are not also found in the south of Italy, and which belong more especially to the Greek flora. There remain 99 species which the Greek floras have not yet announced, spontaneous in Zante; among these, 80 belong to the Italian species; 10 are not cited in Italian floras; and the remaining 9 which at present have only been found in Zante.

It is worthy of observation, that the Leguminosæ and Compositæ are about equal in Zante, as is usually the case in countries in the same latitude. To the south of this line the Leguminosæ seem to acquire the preponderance; to the north the Compositæ have the majority.

CHAPTER V.

CONCLUSION.

HAVING now brought our survey of the various regions to a close, we may cast a glance over the general results of our inquiry, with a view to remark the principal features and contrasts presented by the continent of Europe as a whole.

The great central wall of mountains running in an almost unbroken line from Turkey to the west coast of Spain, forms a natural division between the two very distinct regions of North and South Europe. The latter is a country of mountains, valleys and sea-coast, while in Northern Europe we have a widely expanded tract of almost level land, in the two great adjoining plains of N. Germany and Russia.

Through this mountainous character, the Italian, the Greek and the Spaniard possess great advantages over the North-European; for in accordance with the well-known law of the influence of altitude upon climate, they can ascend from their own southern valleys, full of luxuriant vegetation, to the mountain sides clothed by the rye-fields, the meadows, and the hazel bushes of the north, and seek around the alpine summits the hardy little members of the Lapland flora, or at any rate find there, amid the snow and ice which exist around their peaks through winter and summer, a vegetation which will furnish them with an adequate idea of the scanty alms bestowed by the earth in arctic regions.

But the untravelled northern must be satisfied with hearing of the evergreen woods, the olive groves, the

orange gardens, and the like, which flourish in the clear
air, and bear unscathed the comparatively mild temperature of winters of the South. Yet a contemplation of the
conditions of Northern Europe reveals, that though less
richly endowed, it is not less cared for than the South,
and a multitude of influences are found at work, modifying the law of diminution of temperature with increasing latitude, and producing a variety in the phænomena more than compensating for the deficiency in
those features which have given a romantic celebrity to
the lands of the ancient civilization of Europe.

The greater difference of the seasons, and the comparatively high summer temperature of the North, exercise a very advantageous influence on the vegetation
there; for although the cold of winter arrests the activity of vegetable life, it does not destroy it, and the high
summer heat, in the season of the growth generally, and
in the time of ripening of fruits and seeds in particular,
is exceedingly favourable. If the seasons were equable,
the North would have an eternal spring, snow and ice
would never be seen, for instance in England; but neither
would corn ripen; probably even there would be no
woods, except perhaps in the south-west corner; for at
Quito, in the table-land of Peru, where the seasons are
very equable, the culture of wheat ceases at the mean
temperature of Milan, and the woods disappear at the
mean of Penzance, lower than that of London. This
favouring influence of unlike seasons is seen also in comparing the coasts and inland regions. Iceland and the
Feroës have neither corn nor forests, while both occur on
the mainland in places which have a much lower mean
temperature; the limits of the vine and maize rise higher
towards the north in Germany than on the west coast of

France.   Maize ripens in the valleys of Tyrol, where
snow lies upon the ground during five months of the
year, while it seldom becomes perfectly matured even in
the South of England.

Those plants which require a mild winter will not
grow in the North of Europe, but they advance along
the western coast under the influence of the maritime
climate, and the myrtle of the South is seen in the S.W.
of England.

From the greater difference of the seasons, the approach
of spring is more striking in the North than in the South.
A gentle warmth succeeds to the severe cold of winter,
the lakes and rivers thaw, the snowy covering of the soil
vanishes and gives place to grass and herbs, the trees
and shrubs burst into leaf, the migratory birds return,
and the insect world comes forth from its winter hiding-
places.   In the South, where no snow lies upon the
ground, where the fields and meadows are green through
the winter, and most of the trees and shrubs retain their
leaves, the changes are less important ; merely more
plants grow up and flower, more trees become clothed
with leaves, and animal life shows itself more abundantly.

The annual amount of rain differs so much in different
places, according to exposure to the easterly or westerly
winds, that there is no striking contrast between North
and South in this respect; the South however appears
to have the most.   But the difference is most remarkable in the annual distribution of the rain.   In the North
it is tolerably equable throughout the year, the summer
and autumn having rather a preponderance; in the South,
on the contrary, the summer rains are slight, and autumn
and winter are the proper rainy seasons ; and the winter

rains exceed the summer rains proportionably in going from north to south.

With regard to the wild vegetation, the most striking characters of the South are, the greater variety of plants, especially of trees and shrubs, the more frequent occurrence of tropical forms, the greater abundance of climbing plants, bulbous plants, aromatic herbs and beautiful flowers, and the evergreen woods almost peculiar to those regions. In the North, the fresh green of the meadows and woods in summer greatly surpasses that of the South, where the want of summer rain, and the much higher temperature, give the vegetable world a parched, greyish-yellow aspect at this season of the year.

The warmest parts of Southern Europe appear to be the coast regions of Spain, France and Italy, lying opposite to Africa and bordering the Mediterranean. The mean temperature of Palermo, and even that of Naples, exceeds that of Lisbon. The greatest difference in the seasons in S. Europe occurs in the eastern part of Greece and on the table-land of Spain; it is greater in Italy than on the west coast of Spain.

But as we have seen, climate alone does not determine the specific characters of the vegetation; we have in Europe numerous distinct types, or *floras*, apparently created in particular regions and now undergoing gradual diffusion; and endeavours have been made to trace the course of the diffusion of some of these, which we have already dwelt upon in treating the special regions. A few more words may be added on this highly interesting subject, which, at present in its infancy, is daily attracting a greater amount of attention.

It was intended originally to have given some account of the ancient floras of Europe; of the plants, now buried in a fossil condition in the solid rocks, which once clothed the lands of these latitudes with vegetations of so different a character from that which has formed the subject of our essay. Various reasons, and in particular the length to which the description of existing conditions has extended, induce us to relinquish this intention, which is the less to be regretted, since, in spite of the laborious investigations of many distinguished observers, our knowledge of the real extent of the various fossil floras is as yet exceedingly imperfect, and there still exists much difference of opinion with respect to the determination of their contemporaneous existence.

All these ancient floras have passed away, and no single plant that now exists has been found at all events in any formation earlier than the tertiaries. With regard to those detected in these more recent deposits, they differ from the plants now inhabiting Europe, and are allied to the vegetation now inhabiting other parts of the globe; so that we may place the epoch of the *entrance* of the oldest species now existing on our continent at most no further back than the termination of the eocene period, and many are inclined to assign to the *creation* of modern European plants an equally, if not much more, recent era.

Prof. Schouw has attempted to deduce evidence of the comparative age of the existing floras from the relation of the character of their species to those of fossil floras, of which the comparative antiquity has been pretty certainly determined. In tracing the changes which have taken place in the constituents of the vegetation of different geological periods, it appears as if the plants of successive formations assumed higher types of organization as they were successively created; the earliest plants of which fossil remains have been found seem to have been sea-weeds and cellular plants; after these came the higher Cryptogamous plants, such as Ferns and their allies, with Conifers and Monocotyledons; the Apetalous Dicotyledons seem to have preceded the Polypetalous families, and the Monopetalous families are but sparingly represented until we come to existing forms. M. Brongniart characterizes the three tertiary epochs, the *eocene, miocene,* and *pliocene* of Europe, in the following manner.

1. In the eocene epoch: presence but rarely of Palms, confined to a small number of species; predominance of Algæ and marine Monocotyledons, which must be attributed to the great extent of marine formations during this epoch. The appearance of a great number of extra-European forms, resulting from the presence of the fossil fruits of Sheppey, is an anomaly which is to be explained by the probable existence of marine currents, which brought these fruits from distant countries, as fruits are now brought to the coasts of Ireland and Norway from the equatorial regions of America by the great Atlantic currents; for the special characters of these fruits remove them far from the plants of which the leaves are found in the same geological epoch.

2. In the miocene epoch: abundance of Palms in the majority of localities incontestably belonging to this epoch; the existence of a tolerably large number of extra-European forms, and particularly of the genus *Steinhauera,* which M. Brongniart regards as a Rubiaceous plant allied to the *Morindæ.*

3. In the pliocene epoch: great predominance and variety of Dicotyledons, rarity of Monocotyledons, and

especially the absence of Palms ; but a general analogy of the forms of the plants to those now inhabiting
the temperate regions of Europe, North America, and
Japan.

A remarkable character of the floras of these three
epochs, and one which becomes most striking in the last,
in which the Dicotyledons are most numerous, is the
absence of the largest and most characteristic families of
the Monopetalæ, nothing hitherto found indicating the
existence of Compositæ, Campanulaceæ, Scrophulariaceæ,
Labiatæ, Solanaceæ, Boraginaceæ, &c.

The only Monopetalæ cited in large numbers belong
to the Ericaceæ and Ilicaceæ, with some Sapotaceæ and
Styraceæ, families which contain almost as many Polypetalous as Monopetalous plants. In the miocene flora
alone have been indicated several Apocynaceæ, and the
Rubiaceous genus *Steinhauera* above-mentioned.

Reasoning on these data, Prof. Schouw has instituted
a comparison of the floras of Scandinavia, Germany, and
the Swiss Alps, and comes to the conclusion that the
Alpine flora is the most recent. Thus, comparing the
flora of the Alps with that of Germany and the fossil
floras of Europe, we get the following numerical results :—

	FOSSIL.	RECENT.		
	Before the	After the		
	Chalk.	Chalk.	Germany.	Alps.
Flowerless Plants	·81	·02	·02	·02
Monocotyledons........	·06	·13	·21	·16
Apetalous Dicotyledons..	·12	·45	·08	·04
Petaliferous Dicotyledons.	·01	·40	·69	·78

The Alps therefore have 78 petaliferous Dicotyledons,
Germany only 69 per cent. The ancient floras sub-

sequent to the chalk formations had 40, before that only 1 per cent. On the other hand, the Apetalous Dicotyledons constitute no more than 4 per cent. of the flora of the Alps ; 8 per cent. of that of Germany (7 per cent. if we deduct sea-shore plants) ; while in the tertiary formations they were 45 per cent., and before the chalk 12. With regard to the flowerless plants the quotients are alike, but the proportions to the whole floras of the ancient epochs are exceedingly different. We must not, however, depend exclusively on numerical proportions, but take into account also the predominance of particular groups, especially characterizing the flora of the Alps, and presenting the greatest variety of forms. Of these we find the Ranunculaceæ, Rosaceæ, Saxifrageæ, and Cruciferæ to constitute the largest and most marked forms, while at the same time they belong to the families most developed, and that the next in rank are the Primulaceæ and Gentianaceæ, which may also be considered very prevalent groups. But none of the pentamerous Apetalæ or of the trimerous families are of note in the Alps, much less have they any peculiar form. The alpine plants of these groups are representatives of ordinary German floras.

If we compare in a similar manner the flora of Lapponia, or what amounts nearly to the same thing, the Scandinavian mountain flora, with that of the east of Scandinavia, we obtain the following results :—

	Scandinavia.	Lapponia.
Flowerless	·03	·05
Monocotyledons	·26	·31
Apetalous Dicotyledons	·08	·09
Petaliferous Dicotyledons....	·63	·55

According to geologists, the Scandinavian mountains
are older than the Alps; and yet we find the flora of
Lapland, which is the same as that of the Scandinavian
mountains, approaches more towards fossil vegetation, in-
asmuch as the numerical proportion of flowerless plants is
larger, of Apetalous Dicotyledons somewhat larger, and of
petaliferous Dicotyledons considerably less.   Comparing
together the flora of Lapland, or the Scandinavian mountains, and the Alps, with reference to the numerical extent of their large groups, we perceive a more marked
discrepancy among them, than between the floras of the
Alps and Germany, Scandinavia and Lapland; and yet
if we keep in view the habitual characters of the floras,
their families, genera, and even species, the analogy of
the flora of the Alps with that of the Scandinavian mountains becomes far more manifest than the analogy which
exists between each of them and its corresponding lower
country, and which, according to climate, might be expected to prevail: this becomes evident by combining
the preceding tables :—

	FOSSIL.		RECENT.			
	Before Chalk.	After Chalk.	Ger- many.	Alps.	Scan- dinavia.	Lap- land.
Flowerless	·81	·20	·02	·02	·03	·05
Monocotyledons	·06	·13	·21	·16	·26	·31
Apetalous Dicotyledons	·12	·45	·08	·04	·08	·09
Petaliferous ditto	·01	·40	·69	·78	·63	·55

From these facts Prof. Schouw thinks it reasonable to
conclude that the flora of the Alps is of more recent date
than that of middle Europe or of the Scandinavian mountains, yet he does not consider it as finally proved.

Ingenious as these speculations are, we should join

issue with the author at a previous stage of the inquiry, since he does not accept the hypothesis of special centres of creation (which has been explained in an earlier chapter), considering that migrations could not possibly have produced the intermixtures and widely-diffused conditions of existing species. Consequently his calculations are incompatible with the opinion, that if the same species exists on the Scandinavian mountains and the Alps, some connexion must have existed between these, as the two assemblages of that species have had a common parentage. Looking at the matter in this light, the comparison of the flora of Lapland as a whole, with that of the Alps as a whole, in regard to the question of absolute age, is altogether out of the question.

With regard however to the comparison of the two mountain floras with those of the lowlands around them, some useful conclusions might be drawn in the above manner; although under a somewhat different point of view, since we do not consider it necessary that a flora must have been *created* on the spot where it is now found; and it is quite possible that the vegetation around the mountains may have been in great part derived by migration from regions peopled, at an earlier period, by a vegetation of older type than that of the Alps. The question of absolute age is very difficult, and it is probably safer at present to confine ourselves to the investigation of the order of succession of the various groups of species upon particular tracts, and in so doing it is indispensable to define the limits of the groups or " floras " much more minutely than has been done by Prof. Schouw in the researches above referred to.

Let us turn to the different regions we have examined, and recalling their peculiarities, endeavour to ascertain what are the elements out of which the European flora has been compounded.

In Scandinavia we traced the existence of, probably, seven distinct floras, viz. one group of plants possessed in common with Siberia, another with Iceland and N. America, and perhaps a third indigenous Arctic form; a fourth type consists of the common plants of North Germany; a fifth of those of the highlands of Central Germany; the sixth is represented by a few stragglers from what we have called the West Germanic type; and the seventh, the poorest, the French type.

Britain has all these types except the Siberian; and added to them are the plants of the Pyrenean type in Ireland.

In the North European plains the same types prevail as in the south of Scandinavia, while in the great East European plains the West Germanic and French types are lost, while a new form, that of the "steppe-plants," appears in the south and east.

The Central European highlands have all the types indicated for Britain, with a new flora in the east, consisting of forms from the Danubian provinces, together with stragglers from the Alps, which are not alpino-boreal species.

The Alps possess the common German plants, the alpino-boreal forms in common with Scandinavia and Britain, another proper alpine flora, partly common to the Pyrenees, and on the south side, species derived from northern Italy.

The Pyrenees have some few peculiar alpine forms, but their flora is mostly shared with the countries between which they lie.

Spain possesses in the north and centre a high development of the German and French types, together, in the south, with a peculiar mountain flora extending partly to the Pyrenees, with certain steppe-plants, found also in S. Russia. All the eastern and southern coast provinces have the Mediterranean flora.

Italy appears to have German forms, mingled with a North-Italian type in Lombardy, while peculiar forms occur in Calabria and Sicily. The coasts, together with the Mediterranean shores of France, and the islands, have the great Mediterranean flora, in which the origin of the constituent parts is now probably undistinguishable, although centres of diffusion for certain forms may be noticed in Spain, Italy, Greece, Asia Minor, or the African coast respectively.

Greece also possesses a peculiar flora, probably in a great measure common to Asia Minor and the adjacent regions, but not met with in Italy or Spain. Throughout the south of Europe, however, a great number of the German and French forms are universally distributed, forming thus the basis of the entire European flora, upon which the numerous local variations are superadded.

In this brief summary we have omitted the straggling species of undoubted exotic origin, which have been introduced by extraordinary means, and form mere isolated facts among the general phænomena of distribution. Such are *Erigeron canadense* and *Œnothera biennis*, *Agave americana*, *Mimulus luteus*, &c., all brought by human agency from America.

The most curious points in this distribution are the simultaneous occurrence of the alpino-boreal species on the Scandinavian, British and German mountains, and on the Alps, that of the Pyrenean species in Ireland, and of the S. Russian in Spain; facts which at first sight seem inexplicable, if we accept the doctrine of single centres of diffusion, and indeed to afford plausible arguments for the other hypothesis, that species have been created indifferently wherever the conditions were fitted for their growth.

Yet, in the investigation of phænomena which are the result of combinations of conditions so highly complicated, we must be especially careful to avoid drawing conclusions from too few premises; we must not generalize from the facts which lie near the surface, but must seek to know the influences of the less obvious causes.

The possibility of modification of the vegetation of a country by geological change has already been indicated, both in a general manner in the Introduction, and in a special case referring to the flora of N. Germany. But attention was first drawn to this line of inquiry by Prof. E. Forbes, in his endeavours to trace the origin of the flora and fauna of the British Islands. In these endeavours he sets out with the hypothesis, in our opinion the sound one, of the creation of species in single centres; and proceeding from this assumption, he finds that the present condition of the vegetation and animal life in the British Isles must be explained by the aid of geological facts and reasoning, since the great changes of condition of the face of the islands, and ruptures of previously existing connection with the continent of Europe, which geology demonstrates to have taken place, appear to offer the most satisfactory solution of the question of derivation of the diverse forms by which our islands are now peopled.

Prof. Forbes's views, which rest perhaps more strongly on zoological than botanical facts, were first put forth at the Meeting of the British Association in 1845, and

are laid down more at length and supported by fuller
evidence in the 'Memoirs of the Geological Survey.' It
will of course be impossible for us to enter into the discussion of the many questions here involved, since so
much depends upon the settlement of points of detail
and individual facts: it must suffice to give a general
outline of the theory of the migrations of the different
floras into the British Islands from the Continent.

The British flora is divided by Prof. E. Forbes into
five principal groups; the grouping of the forms differs
slightly from that we have given (p. 168), but the differences are not such as to interfere with the argument.
His groups are:—1. The *Iberian* type, corresponding
to our third group. 2. The *French* type, which corresponds to what we have indicated under this title.
3. The *English* type, which includes most of those forms
comprehended in our *west-* and *mid-Germanic* types.
4. The *Alpine*, our *mountain* plants; and 5. the *Germanic*, which represents our division *a*, the *universal
Germanic* plants. In this subdivision it is not attempted
to enter more into detail than is necessary for the principal features; and in adverting to exceptional cases, the
author states his confident opinion that they are not of
sufficient importance to affect the general argument.

Now, in tracing the organic remains contained in the
more recent geological formations, we meet, in the eocene
tertiary epoch, in the æra of the deposition of the London clay, with decided evidence of the existence at that
time of a flora, most distinct from any other above, upon
those portions of our island then above water. All these
forms are lost, and this flora, now extinct, gives a fixed
point *subsequent* to which the changes under examination must have taken place. Then again, the migra-

tions must all, or at any rate in great part, have taken place *before* the historical period. Between these periods occurred geological changes of the highest importance in reference to the modification of the surface of our islands, its climate, and the distribution of its organic forms.

Geological inquiry has proved that the lowlands of Great Britain are portions of the bed of a pre-existing sea, and that when this bed was elevated the dry land extended over to the north-western parts of Germany, from which our islands again became divided by a subsequent disruption, resulting in the formation of the German Ocean. While these great plains were a seabottom, the mountains which now tower above the surrounding surface could be but islands, projecting much less evidently above the surface of the waves flowing over the 'Germanic plains.' The fossil remains of animals contained in the exposed portions of this ancient seabottom bear testimony that the climate of the northern and part of central Europe was then far colder than it is at present; in fact, that all that part of Europe lying north of a line drawn across the southern part of Ireland and England to the Ural Mountains, was a region characterized by phænomena nearly identical with those now presenting themselves on the N.E. coast of America, within the line of summer-floating ice. It was a region of icebergs and boulders, and the æra of its existence being termed the 'glacial period,' this is known as the glacial sea.

During this epoch it was, as Prof. E. Forbes believes, that our now elevated tracts were peopled by the arctic and alpine species met with upon them at the present day. At that time, low islands in an icy sea, with the means of transport existing in abundance, the mountain chains of N. Europe displayed an arctic flora, which, under the altered climate of our own times, is able to maintain its footing, according to the general rule of climatal influence, through the elevation of the tracts which it inhabits. This flora would probably differ slightly in different parts of its area, and hence part of the variations now existing between the alpine floras of Europe; differences might further result from accidental destruction of the localities of plants scattered sporadically, and from the extinction of forms by various causes during the long period which has elapsed since they first became 'mountain plants.'

The glacial period was succeeded by one presenting a very different condition of things in N. Europe; the bed of the glacial sea was upheaved, the dry land became continuous from Belgium to Britain and from Britain to Ireland, and the former islands and ridges of the icy sea now became peaks and mountain chains. The difference of climate, indicated by fossil evidence, probably depended in a great measure on changes of the distribution of land and water in the Atlantic Ocean, as yet uninvestigated; clearly, however, it would continue to the arctic plants their necessary conditions, elevated as their habitats were to mountain summits. Over the great Germanic plain there appears to have been a migration of both plants and animals from east to west, the descendants of which still constitute the great body of the flora and fauna of the British lowlands. Comparison of Ireland and Britain, and Britain with Belgium, seem to indicate the gradual character of the migration, even in the existing species, which owe their presence in these countries to its occurrence. In this way our islands obtained the plants of Prof. Forbes's 'Germanic type,' corresponding to the universal and mid-Germanic, noticed above (pp. 169, 379).

The 'Kentish flora' (our west-Germanic) evidently passed across the tract once connecting the south of England with the north of France; and Prof. Forbes, although regarding it as a prior invader to the last, considers this question open. Then, to the west, we have the 'French type,' also regarded by Prof. Forbes as an 'older inhabitant' than the Germanic, and even than the 'Kentish.' It constitutes the remains of a flora existing upon a tract of land extending from the northern parts of France, including the Channel Islands, and continued into the S. and S.W. of England, an area now in great part occupied by sea.

It only remains to explain the origin of the Iberian flora of the W. of Ireland. The hypothesis which Prof. Forbes offers to account for this is, that at an epoch anterior to that of any of the other four floras, the west of Ireland was geologically united with the north of Spain; that the flora of the intervening land was a continuation of the flora of the Peninsula; that the northernmost bound of that flora was probably in the line of the western region of Ireland; that the destruction of the intermediate land had taken place before the glacial period; and that during the last-named period climatal changes destroyed the mass of this southern flora remaining in Ireland, the survivors being such species as were most hardy, saxifrages, heaths, &c.

This hypothesis, bold as it appears to be, is supported by a number of very striking facts. There is strong geological probability that a great continent was formed by the upheaval of miocene tertiaries, extending far into the

Atlantic, the coast-line of which it is more than likely that the great belt of gulf-weed now marks, between the 15° and 45° N.L.; and that this tract bore the peculiar fauna and flora now known as Mediterranean, fragments of which are still met with isolated in the Azores, Madeira, and the Canaries, as well as in Ireland and on the Spanish peninsula.

It is urged by some objectors to the above hypothesis, that the existence of like favourable conditions is alone sufficient to account for the co-existence of these plants on these scattered points, or that the ordinary modes of conveyance of seeds, &c., are fully adequate to the colonization of such isolated fragments of an area from one original centre. It appears to us that these views are in the present case equally hypothetical, and by no means more probable than those propounded by Prof. E. Forbes, always provided that the geological evidence derived from tertiary fossils and existing animals is sufficient to demonstrate the facts connected with the glacial period, and the modifications of the area of the Mediterranean fauna and flora. That these data are sound, seems to be borne out by every new fact that comes to light; and it was clearly under a misapprehension of the state of the case in this respect, that Prof. Grisebach laboured when criticising the above views in his 'Annual Report on Geographical Botany.'

Prof. E. Forbes states that he is prepared to extend his views, more especially those relating to the distribution of the alpine plants during the glacial period, to the mountains of Central and Northern Europe. Indeed there does not seem to be any great difficulty in this; but it must be borne in mind that the ridges and islands these mountains at that time constituted, must themselves have possessed truly indigenous species, and thus formed centres of diffusion, which would account for the resemblances and differences of the alpine regions throughout Europe. Whether or not, as Prof. E. Forbes is inclined to assume, the 'Germanic flora' is of more recent creation than the 'Alpine,' and others, is a very difficult question. If we adopt the mode of reasoning used by Schouw, founded on the per-centage of families, in comparison with fossil floras, we should be led to suppose that the 'Germanic type' did already exist before the 'Alpine,' and merely migrated into N. Europe as the face of the country became fitted to receive it. But it seems to us, that reasoning on the per-centage of forms can be no safe guide in the present state of our knowledge of fossil plants, since, for example, the fossil flora of the European pliocene epoch is founded chiefly on species from two localities—Œningen and Parschlug; and the question of the relative age of existing floras must be left over until we know more concerning the real centres of diffusion, and have more distinctly traced the course of migration of forms, especially as dependent on geological changes, which not only modify so importantly the superficial character of regions, but likewise influence, and often over a much wider sphere, those climatal phænomena on which the permanence of the forms of vegetation so much depends.

Coming down from those more ancient times to the æra of the existence of the human race upon the earth, the question arises whether vegetation has undergone any great changes, independently of man's influence, during the historical period. This point must be very briefly discussed, and this is of the less consequence, since there is almost universal agreement among those who have studied these subjects most attentively, that no species has been created since man. The written records left us by the Greeks and Romans indicate that the vegetation of Italy, Greece and Egypt were the same 2000 years ago as at the present day; and this is likewise proved to a certain extent by paintings and actual remnants of vegetables met with in the excavations of Pompeii. If the cultivated plants were the same, the natural conclusion is that the climate differed little; then if the climate remained unaltered, there is no reason why we should expect an alteration in the vegetation; and indeed, the S. European evergreen trees, the laurel, the arbutus, oleander, myrtle, cork oak, ilex, &c., now strongly marking the distinction between N. and S. Europe, are the trees which the classic scholar seeks upon the Mediterranean, as familiar to him through the Greek and Roman authors. Moreover, our northern trees are described by Theophrastus and Pliny as inhabiting, then as now, the more elevated districts.

Prof. Schouw relates an interesting series of observations made by M. Hofman, during thirty years, upon a piece of land reclaimed from the sea in the island of Fuhnen. M. Hofman, being an active botanist, kept a careful journal of the first appearance and subsequent progress of every plant which sprang up spontaneously upon this tract, and it proved that every one of them was a well-known species, belonging to the contiguous floras. M. Hofman assumed that the forms were created upon this tract; and the rapidity with which newly exposed surfaces of land frequently become overgrown, would perhaps lead a superficial observer to this assumption, or at least furnish plausible data to any one desiring to maintain the theory of 'equivocal generation.' But it is unlikely

2 c

such an hypothesis would be adopted very extensively, since so far more simple an explanation is obtained by regarding the new vegetation as a colony from contiguous districts.

The rapidity with which such colonization sometimes takes place, is well illustrated by the enormous diffusion of certain European thistles in South America, to which regions, like the horse, also now spread there in a most wonderful manner in a wild condition, they were carried by European emigrants. Several North American plants have become greatly diffused in Europe during the interval since the discovery of the New World; and we have recently had brought to light a remarkable instance of rapid diffusion in the case of the *Anacharis Alsinastrum,* a water-plant of rapid growth and brittle texture, which is readily broken up into small fragments, each capable of forming an independent plant; these pieces float down running streams and soon fill small watercourses by their vigorous growth. The North American *Impatiens fulva* spreads along water-courses in like manner; while in our land-plants the instances are so numerous, that all our descriptive works and lists contain a large number of species occurring extensively in Britain, but bearing such strong marks of comparatively recent introduction, that many authors consider them as aliens, and refuse them a place in our ' Flora.'

In dismissing this somewhat laborious task, the fruits of which, after all, could in such limited space amount but to an imperfect and fragmentary *résumé* of the subject, the author may be permitted to state, that the present volume is to be regarded merely as a sketch or

rough-draft, in which the principal results of past investigation are for the first time brought together into one
view. The filling up of these outlines, the execution of
a finished picture of the vegetation of Europe, wherein
all the varied phænomena shall be depicted in a clear
light, must be left to a future time, when the study of
details has progressed much further. The author trusts
that these pages may be the means of awakening wider
interest and attracting new labourers into the field.

THE END.

PRINTED BY RICHARD TAYLOR,
RED LION COURT, FLEET STREET.

HISTORY OF ECOLOGY

An Arno Press Collection

Abbe, Cleveland. **A First Report on the Relations Between Climates and Crops.** 1905

Adams, Charles C. **Guide to the Study of Animal Ecology.** 1913

American Plant Ecology, 1897-1917. 1977

Browne, Charles A[lbert]. **A Source Book of Agricultural Chemistry.** 1944

Buffon, [Georges-Louis Leclerc]. **Selections from Natural History, General and Particular, 1780-1785.** Two volumes. 1977

Chapman, Royal N. **Animal Ecology.** 1931

Clements, Frederic E[dward], John E. Weaver and Herbert C. Hanson. **Plant Competition.** 1929

Clements, Frederic Edward. **Research Methods in Ecology.** 1905

Conard, Henry S. **The Background of Plant Ecology.** 1951

Derham, W[illiam]. **Physico-Theology.** 1716

Drude, Oscar. **Handbuch der Pflanzengeographie.** 1890

Early Marine Ecology. 1977

Ecological Investigations of Stephen Alfred Forbes. 1977

Ecological Phytogeography in the Nineteenth Century. 1977

Ecological Studies on Insect Parasitism. 1977

Espinas, Alfred [Victor]. **Des Sociétés Animales.** 1878

Fernow, B[ernhard] E., M. W. Harrington, Cleveland Abbe and George E. Curtis. **Forest Influences.** 1893

Forbes, Edw[ard] and Robert Godwin-Austen. **The Natural History of the European Seas.** 1859

Forbush, Edward H[owe] and Charles H. Fernald. **The Gypsy Moth.** 1896

Forel, F[rançois] A[lphonse]. **La Faune Profonde Des Lacs Suisses.** 1884

Forel, F[rançois] A[lphonse]. **Handbuch der Seenkunde.** 1901

Henfrey, Arthur. **The Vegetation of Europe, Its Conditions and Causes.** 1852

Herrick, Francis Hobart. **Natural History of the American Lobster.** 1911

History of American Ecology. 1977

Howard, L[eland] O[ssian] and W[illiam] F. Fiske. **The Importation into the United States of the Parasites of the Gipsy Moth and the Brown-Tail Moth.** 1911

Humboldt, Al[exander von] and A[imé] Bonpland. **Essai sur la Géographie des Plantes.** 1807

Johnstone, James. **Conditions of Life in the Sea.** 1908

Judd, Sylvester D. **Birds of a Maryland Farm.** 1902

Kofoid, C[harles] A. **The Plankton of the Illinois River, 1894-1899.** 1903

Leeuwenhoek, Antony van. **The Select Works of Antony van Leeuwenhoek.** 1798-99/1807

Limnology in Wisconsin. 1977

Linnaeus, Carl. **Miscellaneous Tracts Relating to Natural History, Husbandry and Physick.** 1762

Linnaeus, Carl. **Select Dissertations from the Amoenitates Academicae.** 1781

Meyen, F[ranz] J[ulius] F. **Outlines of the Geography of Plants.** 1846

Mills, Harlow B. **A Century of Biological Research.** 1958

Müller, Hermann. **The Fertilisation of Flowers.** 1883

Murray, John. **Selections from *Report on the Scientific Results of the Voyage of H.M.S. Challenger During the Years 1872-76.*** 1895

Murray, John and Laurence Pullar. **Bathymetrical Survey of the Scottish Fresh-Water Lochs.** Volume one. 1910

Packard, A[lpheus] S. **The Cave Fauna of North America.** 1888

Pearl, Raymond. **The Biology of Population Growth.** 1925

Phytopathological Classics of the Eighteenth Century. 1977

Phytopathological Classics of the Nineteenth Century. 1977

Pound, Roscoe and Frederic E. Clements. **The Phytogeography of Nebraska.** 1900

Raunkiaer, Christen. **The Life Forms of Plants and Statistical Plant Geography.** 1934

Ray, John. **The Wisdom of God Manifested in the Works of the Creation.** 1717

Réaumur, René Antoine Ferchault de. **The Natural History of Ants.** 1926

Semper, Karl. **Animal Life As Affected by the Natural Conditions of Existence.** 1881

Shelford, Victor E. **Animal Communities in Temperate America.** 1937

Warming Eug[enius]. **Oecology of Plants.** 1909

Watson, Hewett Cottrell. **Selections from *Cybele Britannica.*** 1847/1859

Whetzel, Herbert Hice. **An Outline of the History of Phytopathology.** 1918

Whittaker, Robert H. **Classification of Natural Communities.** 1962